T0281203

The Mathematics of Logic

A guide to completeness theorems and their applications

This textbook covers the key material for a typical first course in logic for undergraduates or first year graduate students, in particular, presenting a full mathematical account of the most important result in logic: the Completeness Theorem for first-order logic.

Looking at a series of interesting systems increasing in complexity, then proving and discussing the Completeness Theorem for each, the author ensures that the number of new concepts to be absorbed at each stage is manageable, whilst providing lively mathematical applications throughout. Unfamiliar terminology is kept to a minimum; no background in formal set-theory is required; and the book contains proofs of all the required set theoretical results.

The reader is taken on a journey starting with König's Lemma, and progressing via order relations, Zorn's Lemma, Boolean algebras, and propositional logic, to Completeness and Compactness of first-order logic. As applications of the work on first-order logic, two final chapters provide introductions to model theory and non-standard analysis.

DR RICHARD KAYE is Senior Lecturer in Pure Mathematics at the University of Birmingham.

The Mathematics of Logic

A guide to completeness theorems and their applications

Richard Kaye

School of Mathematics, University of Birmingham

CAMBRIDGE
UNIVERSITY PRESS

CAMBRIDGE
UNIVERSITY PRESS

University Printing House, Cambridge CB2 8BS, United Kingdom

One Liberty Plaza, 20th Floor, New York, NY 10006, USA

477 Williamstown Road, Port Melbourne, VIC 3207, Australia

314-321, 3rd Floor, Plot 3, Splendor Forum, Jasola District Centre, New Delhi - 110025, India

79 Anson Road, #06-04/06, Singapore 079906

Cambridge University Press is part of the University of Cambridge.

It furthers the University's mission by disseminating knowledge in the pursuit of education, learning and research at the highest international levels of excellence.

www.cambridge.org
Information on this title: www.cambridge.org/9780521708777

© Richard Kaye 2007

First published 2007
Reprinted 2008

A catalogue record for this publication is available from the British Library

ISBN 978-0-521-88219-4 Hardback
ISBN 978-0-521-70877-7 Paperback

Contents

Preface

Mathematical logic has been in existence as a recognised branch of mathematics for over a hundred years. Its methods and theorems have shown their applicability not just to philosophical studies in the foundations of mathematics (perhaps their original *raison d'être*) but also to 'mainstream mathematics' itself, such as the infinitesimal analysis of Abraham Robinson, or the more recent applications of model theory to algebra and algebraic geometry.

Nevertheless, these logical techniques are still regarded as somewhat 'difficult' to teach, and possibly rather unrewarding to the serious mathematician. In part, this is because of the notation and terminology that still survives as a relic of the original reason for the subject, and also because of the off-putting and didactically unnecessary logical precision insisted on by some of the authors of the standard undergraduate textbooks. This is coupled by the professional mathematician's very reasonable distrust of so much emphasis on 'inessential' non-mathematical details when he or she only requires an insight into the mathematics behind it and straightforward statements of the main mathematical results.

This book presents the material usually treated in a first course in logic, but in a way that should appeal to a suspicious mathematician wanting to see some genuine mathematical applications. It is written at a level suitable for an undergraduate, but with additional optional sections at the end of each chapter that contain further material for more advanced or adventurous readers. The core material in this book assumes as prerequisites only: basic knowledge of pure mathematics such as undergraduate algebra and real analysis; an interest in mathematics; and a willingness to discover and learn new mathematical material. The main goal is an understanding of the mathematical content of the Completeness Theorem for first-order logic, including some of its mathematically more interesting applications. The optional sections often require additional background material and more 'mathematical maturity' and go beyond a typical first undergraduate course. They may be of interest to beginning post-graduates and others.

The intended readership of this book is mathematicians of all ages and persuasions, starting at third year undergraduate level. Indeed, the 'unstarred' sections of this book form the basis of a course I have given at Birmingham University for third and fourth year students. Such a reader will want a good grounding in the subject, and a good idea of its scope and applications, but in general does not require a highly detailed and technical treatment.

On the other hand, for a full mathematical appreciation of what the Completeness Theorem has to offer, a detailed discussion of some naive set theory, especially Zorn's Lemma and cardinal arithmetic, is essential, and I make no apology for including these in some detail in this book.

This book is unusual, however, since I do not present the main concepts and goals of first-order logic straight away. Instead, I start by showing what the main mathematical idea of 'a completeness theorem' is, with some illustrations that have real mathematical content. The emphasis is on the content and possible applications of such completeness theorems, and tries to draw on the reader's mathematical knowledge and experience rather than any conception (or misconception) of what 'logic' is.

It seems that 'logic' means many things to different people, from puzzles that can be bought at a newsagent's shop, to syllogisms, arguments using Venn diagrams, all the way to quite sophisticated set theory. To prepare the reader and summarise the idea of a completeness theorem here, I should say a little about how I regard 'logic'.

The principal feature of logic is that it should be about reasoning or deduction, and should attempt to provide rules for valid inferences. If these rules are sufficiently precisely defined (and they should be), they become rules for manipulating strings of symbols on a page. The next stage is to attach meaning to these strings of symbols and try to present mathematical justification for the inference rules. Typically, two separate theorems are presented: the first is a 'Soundness Theorem' that says that *no incorrect deductions* can be made from the inference rules (where 'correct' means in terms of the meanings we are considering); the second is a 'Completeness Theorem' which says that *all correct deductions* that can be expressed in the system can actually be made using a combination of the inference rules provided. Both of these are precise mathematical theorems. Soundness is typically the more straightforward of the two to prove; the proof of completeness is usually much more sophisticated. Typically, it requires mathematical techniques that enable one to create a new mathematical 'structure' which shows that a particular supposed deduction that is not derivable in the system is not in fact correct.

Thus logic is not only about such connectives as 'and' and 'or', though the main systems, including propositional and first-order logic, do have symbols for these connectives. The power of the logical technique for the mathematician arises from the way the formal system of deduction can help organise a complex set of conditions that might be required in a mathematical construction or proof. The Completeness Theorem becomes a very general and powerful way of building interesting mathematical structures. A typical example is the application of first-order logic to construct number systems with infinitesimals that can used rigorously to present real calculus. This is the so-called nonstandard analysis of Abraham Robinson, and is presented in the last chapter of this book.

The mathematical content of completeness and soundness is well illustrated by König's Lemma on infinite finitely branching trees, and in the first chapter I discuss this. This is intended as a warm-up for the more difficult mathematics to come, and is a key example that I refer back to throughout the book.

Zorn's Lemma is essential for all the work in this book. I believe that by final year level, students should be starting to master straightforward applications of Zorn's Lemma. This is the main topic in Chapter 2. I do not shy away from the details, in particular giving a careful proof of Zorn's Lemma for countable posets, though the details of how Zorn's Lemma turns out to be equivalent to the Axiom of Choice is left for an optional section.

The idea of a formal system and derivations is introduced in Chapter 3, with a system based on strings of 0s and 1s that turns out to be closely related to König's Lemma. In the lecture theatre or classroom, I find this chapter to be particularly important and useful material, as it provides essential motivation for the Soundness Theorem. Given a comparatively simple system such as this, and asked whether a particular string σ can be derived from a set of assumptions Σ, students are all too ready to answer 'no' without justification. Where justification is offered, it is often of the kind, 'I tried to find a formal proof and this was my attempt, but it does not work.' So the idea of a careful proof by induction on the length of a formal derivation (and a carefully selected induction hypothesis) can be introduced and discussed without the additional complication of a long list of deduction rules to consider. The idea of semantics, and the Soundness and Completeness Theorems, arises from an investigation of general methods to show that certain derivations are not possible, and, to illustrate their power, König's Lemma is re-derived from the Soundness and Completeness Theorems for this system.

The reader will find systems with mathematically familiar derivations for the first time in Chapter 4. Building on previous material on posets, I develop a system for derivations about a poset, including rules such as 'if $a < b$ and

$b < c$ then $a < c$'. The system also has a way of expressing statements of the form 'a is not less than b', and this is handled using a Reductio Ad Absurdum Rule, a rule that is used throughout the rest of the book. By this stage, it should be clear what the right questions to ask about the system are, and the mathematical significance of the Completeness Theorem (the construction of a suitable partial order on a set) is clear. As a bonus, two pretty applications are provided: that any partial order can be 'linearised'; and that from a set of 'positive' assumptions a 'negative' conclusion can always be strengthened to a 'positive' one.

The material normally found in a more traditional course on mathematical logic starts with Chapter 5. Chapters 5 to 8 discuss boolean algebras and propositional logic. My proof system for propositional logic is chosen to be a form of natural deduction, but set out in a linear form on the page with clearly delineated 'subproofs' rather than a tree structure. This seems to be closest to a student's conception of a proof, and also provides clear proof strategies so that exercises in writing proofs can be given in a helpful and relatively painless way. (I emphasise the word 'relatively'. For most students, this aspect of logic is never painless, but at least the system clearly relates to informal proofs they might have written in other areas of mathematics.) I do not avoid explaining the precise connections between propositional logic and boolean algebra; these are important and elegant ideas, and are accessible to undergraduates who should be able to appreciate the analogies with algebra, especially rings and fields. More advanced students will also appreciate the fact that deep results such as Tychonov's Theorem and Stone Duality are only a few pages extra in an optional section. However, if time is short, the chapter on filters and ideals can be omitted entirely.

Chapters 9 and 10 are the central ones that cover first-order logic and the main Completeness Theorem. Apart from the choice of formal system (a development of the natural deduction system already used for propositional logic) they follow the usual pattern. These chapters are the longest in the book and will be found to be the most challenging so I have deliberately avoided many of the technically tricky issues such as: unique readability; the formal definition of the scope of a quantifier; or when a variable may be substituted by a term. An intelligent reader at this level using his or her basic mathematical training and intuition and following the examples is sure to do the 'right thing' and does not want to be bogged down in formal syntactic details. These technical details are of course important later on if one becomes involved in formalising logic in a first-order system such as set theory or arithmetic. But the place for that sort of work is certainly not a *first* course in logic. For those readers that need it, further details are available on the companion web-pages.

The method of proof of the Completeness Theorem is by 'Henkinising' the language and then using Zorn's Lemma to find a maximal consistent set of sentences. This is easier to describe to first-timers than tree-constructions of sets of consistent sentences with their required inductive properties, but is just as general and applicable. Two bonus optional sections for adventurous students with background in point-set topology include a topological view of the Compactness Theorem, and a proof of the full statement of the Omitting Types Theorem via Baire's Theorem, which is proved where needed.

Chapters 11 and 12 (which are independent of each other) provide applications of first-order logic. Chapter 11 presents an introduction to model theory, including the Löwenheim–Skolem Theorems, and (to put these in context) a short survey of categoricity, including a description of Morley's Theorem. This chapter is where infinite cardinals and cardinal arithmetic are used for the first time, and I carefully state all the required ideas and results before using them. Full proofs of these results are given in an optional section, using Zorn's Lemma only. The traditional options of using ordinals or the well-ordering principle are avoided as being likely to beg more questions than they answer to students without any prior knowledge in formal set theory. Chapter 12 presents an introduction to nonstandard analysis, including a proof of the Peano Existence Theorem on first-order differential equations. My presentation of nonstandard analysis is chosen to illustrate the main results of first-order logic and the interplay between the standard and nonstandard worlds, rather than to be optimal for fast proofs of classical results by nonstandard methods.

I have enjoyed writing this book and teaching from it. The material here is, to my mind, much more exciting and varied than the standard texts I learnt from as an undergraduate, and responses from the students who were given preliminary versions of these notes were good too. I can only hope that you, the reader, will derive a similar amount of pleasure from this book.

How to read this book

Chapters are deliberately kept as short as possible and discuss a single mathematical point. The chapters are divided into sections. The first section of each chapter is essential reading for all. The second section generally contains further applications, examples and exercises to test and expand on material presented in the previous section, and is again essential to read and explore. One or more extra 'starred' sections are then added to provide further commentary on the key material of the chapter and develop the material. These other sections are not essential reading and are intended for more inquisitive, ambitious or advanced readers with the background knowledge required. Chapter 8 may be omitted if time is short, and Chapters 11 and 12 are independent of each other.

Mathematical terminology is standard or explained in the text. Bold face entries in the index refer to definitions in the text; other entries provide further information on the term in question.

Additional material, including some technical definitions that I have chosen to omit in the printed text for the sake of clarity, further exercises, discussion, and some hints or answers to the exercises here, will be found on the companion web-site at http://web.mat.bham.ac.uk/R.W.Kaye/logic.

1
König's Lemma

1.1 Two ways of looking at mathematics

It seems that in mathematics there are sometimes two or more ways of proving the same result. This is often mysterious, and seems to go against the grain, for we often have a deep-down feeling that if we choose the 'right' ideas or definitions, there must be only one 'correct' proof. This feeling that there should be just one way of looking at something is rather similar to Paul Erdős's idea of 'The Book' [1], a vast tome held by God, the SF, in which all the best, most revealing and perfect proofs are written.

Sometimes this mystery can be resolved by analysing the apparently different proofs into their fundamental ideas. It often turns out that, 'underneath the bonnet', there is actually just one key mathematical concept, and two seemingly different arguments are in some sense 'the same'. But sometimes there really are two different approaches to a problem. This should not be disturbing, but should instead be seen as a great opportunity. After all, two approaches to the same idea indicates that there are some new mathematics to be investigated and some new connections to be found and exploited, which hopefully will uncover a wealth of new results.

I shall give a rather simple example of just the sort of situation I have in mind that will be familiar to many readers – one which will be typical of the kind of theorem we will be considering throughout this book.

Consider a binary *tree*. A tree is a diagram (often called a *graph*) with a special *point* or *node* called the *root*, and *lines* or *edges* leaving this node downwards to other nodes. These again may have edges leading to further nodes. The thing that makes this a tree (rather than a more general kind of graph) is that the edges all go downwards from the root, and that means the tree cannot have any *loops* or *cycles*. The tree is a *binary* tree if every node is connected to *at most* two lower nodes. If every node is connected to *exactly*

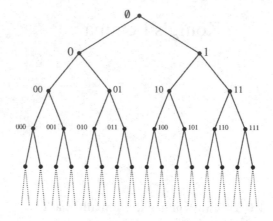

Figure 1.1 The full binary tree.

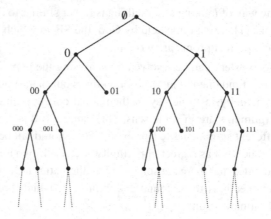

Figure 1.2 A binary tree.

two lower nodes, the tree is called the *full binary tree*. Note that in general, a node in a binary tree may be connected to 0, 1 or 2 lower nodes. We will label the nodes in our trees with sequences of integers. It is convenient to make labels for the nodes below the node that has label x by adding either the digit 0 or 1 to the end of x, giving $x0$ and $x1$. Figure 1.1 illustrates the full binary tree, whereas Figure 1.2 gives a typical (non-full) binary tree.

Trees are very important in mathematics, because many constructions follow trees in some way or other. Binary trees are especially interesting since a *walk* along a tree, following a path that starts at the root, has at most two choices of direction at every node. Binary trees arise quite naturally in many mathematical ideas and proofs and general theorems about them can be quite powerful and useful. One of the better known and more useful of these results is called König's Lemma.

To explain König's Lemma, consider what it means for a tree T to be *infinite*. There are two viewpoints, and two possible definitions.

Firstly, suppose you have somehow drawn the whole of the tree T on paper or on the blackboard and are inspecting it. You are in a fortunate position to be able to take in every one of its features, and to examine every one of its nodes and edges. You will quite naturally say that the tree is infinite if it has infinitely many nodes, or – amounting to the same thing – infinitely many edges. This is a sort of 'definition from perfect information' and is similar to what logicians call semantics, though we will not see the connection with semantics and the theory of 'meaning' for a while.

Now consider you are an ant walking on the binary tree T, which is again drawn in its entirety on paper. You start at the root node, and you follow the edges, like ant tracks, which you hope will take you to something interesting. Unlike the mathematician viewing the tree in its entirety, you can only see the node you are at and the edges leaving it. If you take a walk down the tree, you may have choices of turning left or right at any given node and continuing your path. But it is possible that you have no choice at all, because either there is only one edge out of the node other than the one you entered it by, or possibly there is no such edge at all, in which case your walk has come to an end. To the ant, which cannot perceive the whole of the tree, but just follows paths, there is a quite different idea of what it means for the tree to be infinite: the ant would say that T is infinite if it can find somehow (by guessing the right combination of 'left' and 'right' choices) an infinite path through the tree. The ant's definition of 'infinite' might be thought of as a 'definition from imperfect information' and is similar to the logician's idea of *proof*. If you like, you can think of an infinite path chosen by the ant as a *proof* that the tree is infinite. Like all proofs, it supports the claim made, without giving much extra information – such as what the tree looks like off this path.

König's Lemma is the statement that, for binary trees, these two ideas of a tree being infinite are the same. It is in fact a rather useful statement with many interesting applications. The key feature of this statement is that it relates two definitions, one mathematical definition working from perfect or total information, and one working from the point of view of much more limited information, and shows that they actually say the same thing.

As with all 'if and only if' theorems, there are two directions that must be proved. The first, that if there is an infinite path through the tree then the tree is infinite, is immediate. This easier direction is called a *Soundness Theorem* since it says the ant's perception based on partial information is *sound*, or in other words will not result in erroneous conclusions. The other direction is the non-trivial one, and its mathematical strength lies in the way it states that a rather general mathematical situation (that the tree is infinite) can always be detected in a special way from partial information. The reason why it is called *Completeness* will be discussed later in relation to some other examples.

This has been a long preliminary discussion, but I hope it has proved illuminating. We shall now turn to the more formal mathematical details and define *tree*, *path*, etc., and then state and prove König's Lemma properly.

Definition 1.1 The set of *natural numbers*, $\mathbb{N}$, will be taken in this book to be $\{0, 1, 2, \ldots\}$.

For those readers who expect the natural numbers to start with 1, I can only say that I appreciate that there are occasions when it is convenient to forget about zero, but for me zero is very natural, probably the most logically natural number of all, so is included here in the set of natural numbers.

Definition 1.2 A *sequence* is a function s whose domain is either the set $\mathbb{N}$ of all natural numbers or a subset of it of the form $\{x \in \mathbb{N} : x < n\}$ for some $n \in \mathbb{N}$. Normally the values of the sequence will be numbers, 0 or 1 say, but the definition above (with $n = 0$) allows the empty sequence with no values at all. We write a sequence by listing its values in order, for example as 00110101001 or 0101010101. The *length* of a sequence is the number of elements in the domain of the function. This will always be a natural number or infinity.

Definition 1.3 If s is a sequence of length l and $n \in \mathbb{N}$ is at most l, then $s \restriction n$ denotes the initial part of s of length n.

For example, if $s = 00100011$ then $s \restriction 4 = 0010$.

Definition 1.4 If s is a sequence of length l and x is 0 or 1 then sx is the sequence of length $l + 1$ whose last element is x and all other elements agree with those of s.

Our definition of a tree is of a set of sequences that is closed under the restriction operation $\restriction$.

Definition 1.5 A *tree* is a set of sequences T such that for any $s \in T$ of length n and for any $l < n$ then $s \restriction l \in T$.

Think of a sequence $s \in T$ as a finite path starting from the root and arriving at some node. The individual digits in the sequence determine which choice of edge is made at each node. The set of nodes of the whole tree is then the set of sequences in the set T and two nodes $s, t \in T$ are connected by a single edge if one can be got from the other by adding a single number to the sequence. In other words, s and t are *connected* if $s \restriction (n - 1) = t$ when s is the longer of the two and has length n, or the other way round if t is longer. Then the condition in the definition says, not unreasonably, that each node that this path passes through must also be in the tree. The root of the tree is the empty sequence of length 0.

Definition 1.6 A *subtree* of a tree T is a subset S of T that is a tree in its own right.

A subtree of a tree T might contain fewer nodes, and therefore fewer choices at certain nodes.

Definition 1.7 A *binary tree* is a tree T where all the sequences in it are functions from some $\{n \in \mathbb{N} : n < k\}$ to $\{0, 1\}$.

In other words, at any node, a path from the root of a binary tree has at most two options: to go left (0) or right (1). However, it may turn out that only one, or possibly neither, of these options is available at a particular node.

Definition 1.8 A tree T is *infinite* if it contains infinitely many sequences, or (equivalently) has infinitely many nodes.

A path is a subtree with no branching allowed. That means for any two nodes in the tree, one is a 'predecessor' of the other. More formally, we have the following definition.

Definition 1.9 A *path*, p, in a tree T is a subtree of T such that for any $s, t \in p$ with lengths n, k respectively and $n \leqslant k$, we have $s = t \restriction n$.

A tree T containing an infinite path p is obviously infinite. König's Lemma states that the converse is also true for binary trees.

Theorem 1.10 (König's Lemma) *Let T be an infinite binary tree. Then T contains an infinite path p.*

Proof Suppose T is an infinite binary tree. For a sequence s of length n let T_s be $\{r \in T : r \restriction n = s\} \cup \{s \restriction k : k < n\}$, which we will call the *subtree of T below s*. You will be able to check easily that T_s is a tree. In general it may or may not be infinite.

We are going to find a sequence $s(n)$ of elements of T such that

- $s(n)$ has length n,
- $s(n) = s(n+1) \restriction n$,
- the tree $T_{s(n)}$ below $s(n)$ is infinite.

This construction is by induction, using the third property above as our induction hypothesis. When we have completed the proof the set $\{s(n) : n \in \mathbb{N}\}$ will be our infinite path p in T.

So suppose inductively that we have chosen $s = s(n)$ of length n and T_s is infinite. Then since the tree is binary, made from sequences of 0s and 1s, we have

$$T_s = \{r \in T : r \restriction (n+1) = s0\} \cup \{r \in T : r \restriction (n+1) = s1\} \cup \{s \restriction k : k \leqslant n\}.$$

This is, by the induction hypothesis, infinite. Hence (as the third of these three sets is obviously finite) at least one of the first two sets, corresponding to '0' or '1' respectively, is infinite. If the first of these is infinite we set $s(n+1) = s0$ and in this case we have

$$T_{s(n+1)} = \{r \in T : r \restriction (n+1) = s0\} \cup \{s0\} \cup \{s \restriction k : k \leqslant n\}$$

which is infinite. If not we set $s(n+1) = s1$ which would then be infinite as before. Either way we have defined $s(n+1)$ and proved the induction hypothesis for $n+1$. $\square$

1.2 Examples and exercises

The central theorem of this book, the Completeness Theorem for first-order logic, is not only of the same flavour as König's Lemma, but is in fact a powerful generalisation of it. To give you an idea of the power that this sort of theorem has, it is useful to see a selection of applications of König's Lemma here.

We start by exploring the limits of König's Lemma a little: it turns out that the important thing is not that there are at most two choices at each node but that the number of ways in which the branches divide is always finite.

Definition 1.11 If T is a tree and $s \in T$ is a node of T then the *valency* or *degree* of s is the number of nodes of T connected to s. Thus this is the number of x such that $sx \in T$ plus one (to cater for the edge back towards the root), or just the number of such x if s is the root node.

Exercise 1.12 Prove the following generalisation of König's Lemma: an infinite tree in which every vertex has finite valency has an infinite path. Assume that the tree has vertices or nodes which are sequences of *natural numbers* of finite length and that for each $s \in T$ there is a bound $B_s \in \mathbb{N}$ on the possible values x such that $sx \in T$.

There are two ways that you might have done the last exercise. You might have modified the proof given above, or you may have tried to reduce the case of arbitrary finite valency trees to the case of binary trees by somehow 'encoding' arbitrary finite branching by a series of binary branches.

Exercise 1.13 Whichever method you used, have a go at proving the extension of König's Lemma by the other method.

Exercise 1.14 By giving an appropriate example of an infinite tree, show that König's Lemma is false for graphs with vertices of infinite valency.

König's Lemma is an elegant but nevertheless not very surprising or difficult result to see. Its truth, it seems, is reasonably clear, though a completely rigorous proof takes a moment or two to come up with. It is all the more surprising, therefore that there should be non-trivial applications. We will look at a few of these now, though nothing later in this book will depend on them.

Example 1.15 The set $X = [0, 1]$ has the property (called *sequential compactness*, equivalent to compactness for metric spaces) that every sequence (a_n) of elements of X has a subsequence converging to some element in X.

Proof Starting with $[0, 1]$ we continually divide intervals into equal halves, but at stage k of the construction we throw away any such interval that contains none of the a_n with $n > k$. More precisely, the nodes of the tree at depth k are identified with intervals $I = [(r-1)2^{-k}, r2^{-k}]$ for which $r \in \mathbb{N}$ and $\{a_n : n > k \text{ and } a_n \in I\}$ is non-empty, and two nodes are connected if one is a subset of the other.

This defines a binary tree. It is infinite because there are infinitely many a_n and each lies in an interval. By König's Lemma there is an infinite path through this tree, and by the construction of the tree we may take an infinite subsequence of a_n in this path, one at each level of the tree. This is the required convergent subsequence. □

Now consider infinite sequences $u_0 u_1 u_2 \ldots$ of the digits $0, 1, 2, \ldots, k - 1$. We will call such sequences *k-sequences*. Say a *k*-sequence *s* is x^n-*free* if there is no finite sequence, *x*, of digits $0, 1, 2, \ldots, k - 1$, such that the finite sequence x^n (defined to be the result of repeating and concatenating *x* as $xxxx \ldots x$, where there are *n* copies of the string *x*) does not appear as a contiguous block of the sequence *s*.

Exercise 1.16 (a) Show that there is no x^2-free 2-sequence.

(b) Use König's Lemma to show that there is an x^3-free 2-sequence if and only if there are arbitrarily long finite x^3-free 2-sequences. State and prove a similar result for x^2-free 3-sequences.

(c) Define an operation on finite 2-sequences σ such that $\sigma(0) = 01$, $\sigma(1) = 10$, and $\sigma(u_0 u_1 \ldots u_m) = \sigma(u_0)\sigma(u_1) \ldots \sigma(u_m)$, where this is concatenation of sequences. Let $\sigma^n(s) = \sigma(\sigma(\ldots(\sigma(s))\ldots))$, i.e. σ iterated *n* times. Show that each of the finite sequences $\sigma^n(0)$ is x^3-free, and hence there is an infinite x^3-free 2-sequence.

(d) Show there is an x^2-free 3-sequence.

Another example of the use of König's Lemma is for graphs in the plane. A *graph* is a set *V* of vertices and a set *E* of edges, which are unordered subsets of *V* with exactly two vertices in each edge. In a *planar graph* the set of vertices *V* is a set of points of $\mathbb{R}^2$, and the edges joining vertices are lines which are 'smooth' (formed from finitely many straight-line segments) and may not cross except at a vertex.

A graph with set of vertices *V* can be *k-coloured* if there is a map $f \colon V \to \{0, 1, \ldots, k - 1\}$ such that $f(u) \neq f(v)$ for all vertices *u*, *v* that are joined by an edge. You should think of the values $0, 1, \ldots, k - 1$ as 'colours' of the vertices; the condition says two adjacent vertices must be coloured differently. Graph colourings, and especially colourings of planar graphs, are particularly interesting and have a long history [12]. A deep and difficult result by Appel and Haken shows that every finite planar graph is 4-colourable [10].

Exercise 1.17 Use König's Lemma to show that an infinite graph can be *k*-coloured if and only if every finite subgraph of it can be so coloured. (Make the simplification that the vertices of our infinite graph can be ordered as $v_0, v_1, \ldots$ with indices from $\mathbb{N}$. Construct a tree where the nodes at level *n* are all *k*-colourings of the subgraph with vertices $v_0, v_1, \ldots, v_{n-1}$, and edges join nodes if one colouring extends another.) Deduce from Appel and Haken's result that every infinite planar graph can be 4-coloured.

Tiling problems provide another nice application of König's Lemma. Consider a finite set of *tiles* which are square, with special links like jigsaw pieces so that in a tiling with tiles fitting together, one edge of one tile must be next to one of certain edges of other tiles. A tiling of the plane is a tiling using any number of tiles of each of the finitely many types, so that the whole of the plane is covered. *Tiling problems* ask whether the plane can or cannot be tiled using a particular set.

Exercise 1.18 Prove that a finite set of tiles can tile the plane if and only if every finite portion of the plane can be so tiled.

Finally, for this section, trees are also useful for describing computations. We will not define any idealised computer here, nor provide any background in computability theory, so this next example is for readers with such background, or who are willing to suspend judgement until they have such background. Normally, computations are deterministic, that is every step is determined completely by the state of the machine. A *non-deterministic computation* is one where the computer has a fixed number, B, of possible 'next moves' at any stage. The machine is allowed to choose one of these 'at random', or by making a 'lucky guess' and in so doing it hopes to verify that some assertion is true. This gives rise to a computation tree of all possible computations. Suppose we somehow know in advance that whatever choices are made at any step, every computation of the machine will eventually halt and give an answer. That means that all paths through the computation tree are finite. Then by the contrapositive of König's Lemma the tree is finite. This means that the non deterministic computation can be simulated in finite time by a deterministic one which constructs the computation tree in memory and analyses it.

1.3 König's Lemma and reverse mathematics*

König's Lemma is rather attractive and has some pretty applications. It has been 'traditional' in logic textbooks to give some of the examples above as applications of the much more powerful 'Completeness Theorem for first-order logic'. Whilst not incorrect, this has always seemed a pity to me, as it hardly does the Completeness Theorem justice when the applications can be proved directly from the more familiar König's Lemma. Suffice it to say for now that there will be plenty of interesting applications of the full Completeness Theorem that cannot be argued from König's Lemma alone.

It may be a good idea to say a few words about why König's Lemma is powerful, and where it does real mathematical work. The reason is that, although there may be an infinite path in a tree, it is not always clear how to find one, and in any case there are likely to be choices involved. In our proof of König's Lemma, to keep track of all these individual choices, we used the concept of a certain subtree T_s being infinite. Being 'infinite' is of course a powerful mathematical property, and one about which there is a lot that can be said, both within and outside the field of mathematical logic. This concept of an infinite subtree is doing quite a lot of work for us here, especially as it is being used infinitely many times in the course of an induction.

Some workers in the logic community study these ideas in more detail by trying to identify which theorems need which lemmas to prove them. This area of logic is often called *reverse mathematics* since the main aim is usually to prove the *axioms* from the *theorems*. I am not going to advocate reverse mathematics here, but there are plenty of times when it is nice to know that a complicated lemma cannot be avoided in a proof. It is certainly true for many of the exercises in the previous section that König's Lemma (or something very much like it) is necessary for their solution. In reverse mathematics one usually works from a weaker set of axioms, one where the concept of an infinite set is not available. It turns out, for example, that relative to this weak set of axioms the sequential compactness of $[0, 1]$ is actually equivalent to König's Lemma. For more information on reverse mathematics see the publications by Harvey Friedman, Stephen Simpson and others, in particular Simpson's 2001 volume [11].

The proof of König's Lemma works, as we have seen, by making a series of choices. The issue of making choices is also a very subtle one, but one that will come up in many places in this book. We can always make finitely many choices as part of a proof, by just listing them. (In this way, to make n choices in a proof you will typically need at least n lines of proof, for each $n \in \mathbb{N}$.) But making infinitely many choices in one proof, or even an unknown finite number of choices, will depend on being able to give a rule stating which choice is to be made and when. This might be more difficult to achieve. Some versions of König's Lemma do indeed involve infinitely many arbitrary choices as we turn 'left' or 'right' following an infinite path. This is a theme that will be taken up in the next chapter. As a taster, you could attempt the following exercise, a more difficult version of Exercise 1.12.

Exercise 1.19 Consider the generalisation of König's Lemma that says that an infinite tree T in which every vertex has finite valency has an infinite path. Do not make any simplifying assumptions on the elements of the sequences $s \in T$. What choices have to be made in the course of the proof, and how might you specify all of these choices unambiguously in your proof?

2

Posets and maximal elements

2.1 Introduction to order

The idea of an *order* is central to many kinds of mathematics. The real numbers are familiarly ordered as a number-line, and even a collection of sets will be seen to be *partially* ordered by the 'subset of' relation. We shall start by presenting the axioms for a partially ordered set and then discuss one particularly interesting question about such sets, whether they have *maximal elements*.

An order relation is a relation R between elements x, y of some set X, where xRy means x is smaller than or comes before y. An alternative notation arises when one thinks of the relation more concretely as a set of pairs (x, y), a subset of $X^2 = \{(x, y) : x, y \in X\}$. We can then write xRy in an alternative way as $(x, y) \in R$.

Definition 2.1 A *partial order* on a set X is a relation $R \subseteq X^2$ such that

   (i) $(x, y) \in R$ and $(y, z) \in R$ implies $(x, z) \in R$
   (ii) $(x, x) \notin R$

for all $x, y \in X$.

Example 2.2 The relation on the set of real numbers $\mathbb{R}$ defined by '$(x, y) \in R$ if and only if $x < y$' is a partial order, where $<$ is the usual order on the set of real numbers. In fact it is a special kind of partial order that we will later call a *total order* or *linear order*.

Example 2.3 If X is any set and P is its *power set*, i.e. the set of all subsets of X, then the relation R on P given by $(A, B) \in R$ if and only if A is a proper subset of B (i.e. $A \subseteq B$ and $A \neq B$) is also a partial order on P.

We usually use a symbol such as $<$, etc., for our relation R and when we do we write $x < y$ instead of $(x, y) \in R$.

11

There is another kind of partial order relation corresponding to $\leqslant$ instead of $<$. Here, the order relation is allowed to relate equal elements. In other words, we will allow $(x, x) \in R$ to be true. (This was explicitly disallowed for the first kind of order as defined above.) Clearly we will need different axioms for such an order relation, and the axioms we choose are as follows.

Definition 2.4 A *non-strict partial order* on a set X is a binary relation $R \subseteq X^2$ such that

   (i) $(x, y) \in R$ and $(y, z) \in R$ implies $(x, z) \in R$

  (ii) $(x, x) \in R$

 (iii) $(x, y) \in R$ and $(y, x) \in R$ implies $x = y$

for all $x, y \in X$.

We usually use a symbol such as $\leqslant$, etc., for a non-strict partial order. The first kind of partial order is sometimes called a *strict* partial order to distinguish it from the non-strict case.

If X is any set and P is its power set, then the relation S on P given by $(A, B) \in S$ if and only if A is a subset of B is a non-strict partial order on P because of the law that says two sets A and B are the same if and only if $A \subseteq B$ and $B \subseteq A$.

If $<$ is a strict partial order on X then we can turn it into a non-strict partial order $\leqslant$ by defining $x \leqslant y$ if and only if $x < y$ or $x = y$. In the other direction, given a non-strict partial order $\leqslant$ on X then we can define a strict partial order by $x < y$ if and only if $x \leqslant y$ and $x \neq y$. If we start with one kind of order and do both of these processes, we get back to the original order relation, so strict and non-strict partial orders are versions of the same idea. Some of the exercises discuss these points further.

Definition 2.5 A *poset* is a non-empty set X with a partial order of either the strict or non-strict variety.

The word poset is an abbreviation for *Partially Ordered SET*. Unless a different notation for the order relation is given, we shall use $<$ for the strict partial order on X and $\leqslant$ for its non-strict version.

Example 2.6 Let T be a binary tree, i.e. a set of finite sequences of 0s and 1s which is closed under taking initial segments. Define $\sigma \leqslant \tau$ to mean that σ is an initial segment of τ. Then this is a (non-strict) partial order on T and makes T into a poset.

There will be several occasions throughout this book where we will be interested in the notion of a maximal element in a poset. This is defined next.

Definition 2.7 If X is a poset and $x \in X$, we say that x is a *maximal element* of X if there is no y in x such that $x < y$.

Be careful how you read this. The element x is maximal if there is nothing bigger. This is not the same as saying that x is the biggest element. Indeed a poset may have several maximal elements, as we will see, but there can obviously only be one biggest element.

Example 2.8 In the poset P of the set of all subsets of X, with (non-strict) order relation $\subseteq$, there is a maximal element, X itself. This is because if $A \in P$ then $A \subseteq X$ so we cannot have $X \subseteq A$ and $X \neq A$. In fact X is the only maximal element since $A \subseteq X$ for all $A \in P$, so no $A \neq X$ can also be maximal.

Example 2.9 In the poset $\mathbb{R}$ of the reals with the usual ordering there are no maximal elements.

Example 2.10 Let $X = \{1, 2, 3\}$, and let Q be the set of all subsets of X with at most two elements. This is a poset with the same $\subseteq$ ordering. In this case there are three distinct maximal elements, $\{2, 3\}$, $\{1, 3\}$, and $\{1, 2\}$.

As we have seen in the examples, some posets have a unique maximal element, some have many, and some have none. We are going to discuss some criteria that can be used to determine whether a particular poset has a maximal element. First, we say a poset is finite if its underlying set X is finite. Then we have the following theorem.

Theorem 2.11 *Any finite poset has at least one maximal element.*

Proof This is another proof that involves making choices, somewhat similar to those in the last chapter. Fortunately, only finitely many are required in this case. Let X be our finite poset and let $a \in X$. Let $a_0 = a$. We are going to define a sequence of elements $a_n \in X$ such that $a_n < a_{n+1}$ for each n.

Given $a_n \in X$, there are two possibilities. Either a_n is maximal, in which case we are finished, or else there is some $b \in X$ with $a_n < b$. In the latter case, choose a_{n+1} to be such a b.

Now the argument in the previous paragraph cannot give an infinite sequence of elements of X. This is because X is finite and the sequence a_n is strictly increasing, $a_n < a_{n+1}$, and hence by the axioms for a strict partial order, all the elements a_n are different. Therefore we cannot always have the second option in the previous paragraph, so at some point the a_n that we have obtained will be maximal. That is, our finite poset has a maximal element, as required. □

It may be useful to know when a poset has a biggest, or unique maximal element. Say that a poset X is *directed* if for any two elements a, b in X there is some $c \in x$ such that $a \leqslant c$ and $b \leqslant c$.

Theorem 2.12 *Let X be a directed poset. Then if X has at least one maximal element it has exactly one maximal element.*

Proof Exercise. □

This last result will not be needed in the sequel. Directed posets are occasionally useful, but we are more intereested here in finding a condition for when a poset has *some* maximal element. We have seen that finite sets always have maximal elements. This is not true in general, as in the example of $\mathbb{R}$. (Incidently, $\mathbb{R}$ is directed, so this property does not guarantee the existence of maximal elements either.) We will next look at those sets that are 'only just a bit bigger than finite', the countable sets.

Definition 2.13 A set X is said to be *countable* if either it is empty or else it is of the form $\{a_n : n \in \mathbb{N}\}$ for some sequence of elements a_n from X.

Thus the elements of a countable set can be counted off as $a_0, a_1, a_2, \ldots$ (possibly with repetitions). If you think of $n \mapsto a_n$ as a function $\mathbb{N} \to X$, this means that a non-empty set X is countable if and only if X is empty or there is a surjection from $\mathbb{N}$ onto X.

In particular the next proposition follows directly from this.

Proposition 2.14 *Any finite set is countable. The set of natural numbers $\mathbb{N}$ is countable.*

The next theorem shows that the notion of 'countable sets' has some real substance. It is due to Georg Cantor who 'invented' set theory towards the end of the nineteenth century, and is not quite so straightforward. See Exercise 2.28 for hints on how to prove this result.

Theorem 2.15 *The set of real numbers $\mathbb{R}$ is not countable.*

To explore what properties a poset must have in order to contain maximal elements, consider the set of rational numbers ordered in the usual way. This, like the set of reals, has no maximal element. It also happens to be countable. (See the exercises.) The reason it fails to have a maximal element is that it is 'all lined up' nicely, or as we shall say, it is *linearly ordered*, or is a *chain*, but this chain has no *upper bound*.

Definition 2.16 Let X be a poset. A subset $Y \subseteq X$ is a *chain* if

- for all $x, y \in Y$, either $x < y$ or $y < x$ or $x = y$.

If this holds for all $x, y \in X$ (i.e. if the whole of X is a chain) then we say the poset X is *linearly* or *totally* ordered.

If we are to find maximal elements in a poset X we should at least be able to deal with chains somehow. The next definition gives a possible way that this might be done.

Definition 2.17 Let X be a poset. We say that X has the *Zorn property* if for every chain $Y \subseteq X$ there is an *upper bound* $x \in X$ of Y. That is, there is $x \in X$ such that $y \leqslant x$ for all $y \in Y$.

Note particularly that the element x need not be in Y itself. Sometimes it will be, sometimes not. It must be in X though.

Here is the big theorem of this chapter. The proof is a more sophisticated version of the proof that all finite posets have maximal elements.

Theorem 2.18 *Let X be a countable poset with the Zorn property. Then X has a maximal element.*

Proof Let $X = \{x_n : n \in \mathbb{N}\}$ be our countable poset. We try to repeat the main idea of the previous result on finite posets, constructing an increasing sequence of elements $a_n \in X$ from X.

Take $a_0 = x_0$, the 'first' element in X. We will construct our sequence by making 'choices' as before, but this time we will have infinitely many choices to make so will need to specify how the choices are to be made carefully. At each stage we will have chosen $a_n \in X$. Because a_n is in X it is equal to some x_k, and (in case there is more than one index k for which this is true) let us choose the least k for which $a_n = x_k$. Inductively we will assume that a_n is maximal in $\{x_0, \ldots, x_k\}$ *for this least* k.

If a_n is maximal in the whole of X we are finished, having got what we set out to prove. Otherwise there is $x \in X$ such that $a_n < x$. We must choose one. Do this by choosing $a_{n+1} = x_m \in x$ where $m \in \mathbb{N}$ is the least natural number such that $a_n < x_m$. This m must be greater than k because a_n is maximal in $\{x_0, \ldots, x_k\}$. By our choice of x_m, it must also be maximal in $\{x_0, \ldots, x_m\}$.

Continuing in this way we get an increasing sequence a_n. If we never find any maximal elements, this sequence must be infinite in length and all elements in it distinct. But this would contradict the Zorn property.

To see this, observe first that the set $Y = \{a_n : n \in \mathbb{N}\}$ is a chain in X. The Zorn property implies that it has an upper bound $z \in X$, and by the countability of X we have $z = x_m$ for some m. However, since the sequence a_n has infinite length there is some element in the sequence equal to x_l for some $l \geqslant m$. This quickly gives a contradiction, as x_l is maximal in $\{x_0, \ldots, x_l\}$ so it cannot be that $x_l < x_m$. On the other hand, x_m is an upper bound for Y, so $a_n = x_l < x_m$. $\qquad\square$

As mentioned, the proof of this theorem goes by choosing elements of X repeatedly. The proof works because we had a good recipe for making such a choice – we chose $x_m \in X$ where $m \in \mathbb{N}$ was least possible each time. It seems entirely reasonable that the theorem should be true even if X is not countable, but we would need a different way to make our choices – in fact a new mathematical 'choice principle'. The statement we would like to prove is the following.

Theorem 2.19 (Zorn's Lemma) *Let X be a poset with the Zorn property. Then X has a maximal element.*

There is just such a choice principle, called (reasonably enough) the *Axiom of Choice* that enables us to prove Zorn's Lemma. The Axiom of Choice is one of the usual axioms for set theory introduced in the 1920s by Zermelo, Fränkel and others, and has been widely accepted. It turns out that not only does the Axiom of Choice suffice to prove Zorn's Lemma, but also that the converse is true: from Zorn's Lemma we can prove the Axiom of Choice. Because of this, I will adopt the same approach as many other writers and accept Zorn's Lemma as an extra axiom for set theory, and assume it true and use it whenever necessary. For those who really want to know the details, we present the Axiom of Choice and the proof that it implies Zorn's Lemma in the optional Section 2.3 of this chapter.

Finally, we note a technical but rather general method for showing that the Zorn property holds, one that will apply to almost all examples in this book, including those given as illustrations in the next section. Here we consider Zorn's Lemma in the case of a poset X of subsets of another set B, where the order relation is $\subseteq$.

Proposition 2.20 *Let X be a non-empty poset of subsets $A \subseteq S$ having some property $\Phi(A)$, where the order relation on X is $\subseteq$, i.e.*

$$X = \{A \subseteq S : \Phi(A)\}.$$

Suppose also that the property $\Phi(A)$ defining X is such that

- *if $\Phi(A)$ is false then there are finitely many $a_1, a_2, \ldots, a_n \in A$ such that every $A' \supseteq \{a_1, a_2, \ldots, a_n\}$ fails to satisfy $\Phi(A')$.*

Then X has the Zorn property.

Proof Let $Y \subseteq X$ be a chain, and let $A = \bigcup Y = \{x \in A : A \in Y\}$. Then $A \subseteq C$ for each $C \in Y$. So it suffices to show that A has the property $\Phi(A)$. If not, there are $a_1, a_2, \ldots, a_n \in A$ such that every $A' \supseteq \{a_1, a_2, \ldots, a_n\}$ fails to satisfy $\Phi(A')$. In particular there are $C_i \in Y$ such that $a_i \in C_i$ for each i. But Y is a chain under $\subseteq$, so some C_i must contain all the others. That means $C_i \supseteq \{a_1, a_2, \ldots, a_n\}$ and hence C_i fails to satisfy $\Phi(C_i)$ and thus $C_i \notin X$, a contradiction. □

2.2 Examples and exercises

Exercise 2.21 Suppose X is a poset with ordering $<$ and suppose $Y \subseteq X$ is non-empty. Then $<$ can also be regarded as a relation on Y, and Y is therefore also a poset. What is it about the axioms for a poset that ensure that this is the case? What other cases can you think of where a non-empty subset of the domain of some mathematical structure you have studied is automatically a 'sub-object'? And what about cases where this is not true?

Exercise 2.22 If $<$ is a strict partial order on X then turn it into a non-strict partial order $\leqslant$ as described in the text, and then turn that into a strict partial order $<'$. Show that $<$ and $<'$ are the same.

Do the same exercise, this time starting from a non-strict $\leqslant$, getting the corresponding $<$ and from this a strict $\leqslant'$, and showing that $\leqslant$ and $\leqslant'$ are the same.

To understand the next definition and exercise, it would be instructive to check your argument for the last exercise and identify where the third axiom of a (strict) partial order is required.

Definition 2.23 A *preorder* is a binary relation on a set X satisfying the first two axioms for a non-strict partial order.

Exercise 2.24 Let X be a non-empty set with a preorder $\leqslant$. Define an equivalence relation $\sim$ on X by $x \sim y$ if and only if $x \leqslant y$ and $y \leqslant x$. (You have to prove this is an equivalence relation.) Let $X/\sim$ denote the set of equivalence classes and define $[x] \leqslant [y]$ if and only if $x \leqslant y$, on equivalence classes $[x]$ and $[y]$. Show that this is well defined (i.e. the definition does not depend on the choice of the representatives x, y of the equivalence classes $[x]$ and $[y]$) and defines a (non-strict) partial order on $x/\sim$.

Exercise 2.25 Prove Theorem 2.12.

Exercise 2.26 Prove that the set $\mathbb{Q}$ of rational numbers is countable. (Hint: first show that $\mathbb{Z} \times (\mathbb{Z} \setminus \{0\})$ is countable, and then compose functions.)

The next exercise is often referred to as 'a countable union of countable sets is countable'. It is not quite straightforward how to state it, as some versions of the result require the Axiom of Choice and others do not. The following is a version which does not require the Axiom of Choice.

Exercise 2.27 Suppose that X_i is a countable set for each $i \in \mathbb{N}$ and that there is a function f with domain $\mathbb{N} \times \mathbb{N}$ such that $X_i = \{f(i, j) : j \in \mathbb{N}\}$ for each i. Show that

$$\bigcup \{X_i : i \in \mathbb{N}\} = \{x : x \in X_i \text{ for some } i \in \mathbb{N}\}$$

is countable.

Exercise 2.28 Prove that the set of real numbers is not countable. (Hint: suppose $\mathbb{R}$ is countable and that r_n is a sequence of reals in which every real number appears at least once. Imagine writing down each r_n in decimal form and construct a number $s \in \mathbb{R}$ that differs from each r_n at the nth decimal place. Conclude s is not anywhere in the sequence r_n.)

Exercise 2.29 Let X be any set. Show that the power set P of X is not in one-to-one correspondence with X, i.e. there is no bijection $f: X \to P$, and hence deduce that the power set of $\mathbb{N}$ is not countable. (Hint: consider the set $\{x \in X : x \notin f(x)\}$. Show that this cannot be $f(y)$ for any $y \in X$.)

There are many applications of Zorn's Lemma to algebra. For example, in group theory, Zorn's Lemma can be used to show the following result about subgroups and transversals.

Proposition 2.30 *Let G be a group and H a subgroup of G. Then there is a* *transversal of H in G, i.e. a set $T \subseteq G$ such that for each $g \in G$, $g = ht$ for exactly one pair of $h \in H$ and $t \in T$.*

Proof Let X be the poset of sets T such that

$$h_1 t_1 = h_2 t_2 \text{ implies } h_1 = h_2 \text{ and } t_1 = t_2$$

for all $t_1, t_2 \in T$ and all $h_1, h_2 \in H$. For the order relation on X we take the usual subset-of relation $\subseteq$.

The poset X has the Zorn property since if $Y \subseteq X$ is a chain and

$$S = \bigcup Y = \{x \in T : T \in Y\}$$

then clearly $T \subseteq S$ for each $S \in Y$. We claim that $S \in X$. If not there are $h_1, h_2 \in H$ and $t_1, t_2 \in S$ such that $h_1 t_1 = h_2 t_2$ and $h_1 \neq h_2$ or $t_1 \neq t_2$. But then $t_1 \in T_1 \in Y$ and $t_2 \in T_2 \in Y$ for some T_1, T_2, and as Y is a chain either $T_1 \subseteq T_2$ or $T_2 \subseteq T_1$. Assuming $T_1 \subseteq T_2$ we have $t_1, t_2 \in T_2$ and hence $h_1 t_1 = h_2 t_2$ shows $T_2 \notin X$, which is impossible. $T_2 \subseteq T_1$ is similar.

By Zorn's Lemma, X has a maximal element T. It suffices to show that every $g \in G$ is ht for one pair $h \in H$ and $t \in T$. If not, suppose $g \in G$ is not of the above form. Then $g \notin T$ since $1 \in H$ and if $g \in T$ then $g = g1$, and $T \cup \{g\} \in X$. This last is because if $h_1 t = h_2 g$ for $h_1, h_2 \in H$ and $t \in T$ then $g = h_2^{-1} h_1 t$ writes g as ht for $h = h_2^{-1} h_1 \in H$. But this contradicts the maximality of T and hence there is no such g, as required. $\qquad \square$

Instead of proving the Zorn property directly in the above proof, Proposition 2.20 might have been used.

Exercise 2.31 Prove that the poset X in Proposition 2.30 has the Zorn property by using Proposition 2.20.

A linearly independent subset of a vector space V is one for which no non-trivial *finite* linear combination is zero. A basis of a vector space V is a linearly independent set B such that each $v \in V$ is a linear combination of *finitely* many elements of B. In general there is no way of defining infinite sums in a vector space, so the use of the word 'finite' here is necessary. In other words the definition just given is the 'correct' one, but you may not be used to this emphasis on finiteness. (A typical first course in linear algebra normally deals with finite dimensional spaces only, where this emphasis is unnecessary. For general finite and infinite dimensional spaces, Zorn's Lemma is required in a number of places.) However, it is precisely this finiteness of linear combinations that allows us to apply Proposition 2.20 in the next exercise.

Exercise 2.32 Let V be a vector space over a field F. Show that V has a basis. (Hint: let X be the poset of all linearly independent subsets of V, ordered by the usual set inclusion, $\subseteq$. Explain why X has a maximal element, and then show that a maximal element of X is in fact a basis.)

Exercise 2.33 Let V, W be vector spaces over a field F with bases $B \subseteq V$ and $C \subseteq W$. Suppose there is a bijection $B \to C$. Show that V, W are isomorphic.

Another popular application of Zorn's Lemma to algebra is to find maximal ideals in a ring.

Exercise 2.34 Let I be a non-trivial ideal in a commutative ring R. Show that I extends to a maximal non-trivial ideal M. Show that R/M is a field.

2.3 Zorn's Lemma and the Axiom of Choice*

As indicated, Zorn's Lemma is a version of a set theoretic principle called the *Axiom of Choice*.

Lemma 2.35 (Axiom of Choice) *If X is a set of non-empty sets then there is a function $f\colon X \to \bigcup X$ such that $f(x) \in x$ for all $x \in X$.*

In other words, given a collection of 'choices' to be made (one for each $x \in X$) there is a function – a mathematical object in the realm of set theory – that makes one choice for each simultaneously. This function f is often called a choice function. The set $\bigcup X$ here is simply the set $\{x : x \in y$ for some $y \in X\}$ of all possible elements-of-elements of X.

As mentioned already, we can make any finite number of choices in a mathematical proof. The Axiom of Choice allows us to make an *unbounded number* of choices, or indeed *infinitely many* choices in a single proof. Although it may not be obvious from a rapid inspection of our proof above of Theorem 2.11, this theorem, that any finite poset has a maximal element, does *not* require any form of the Axiom of Choice. To see this, recall that by definition a set X is finite if and only if there is a bijection $f\colon X \to \{0, 1, \ldots, n-1\}$ for some $n \in \mathbb{N}$. This enables us to define a choice function $F\colon P_0 \to X$ where P_0 is the set of non-empty subsets of X, by setting $F(A)$ to be the element $a \in A$ for which $f(a) \in \{0, 1, \ldots, n-1\}$ is least. In particular this definition does not require the Axiom of Choice. This enables all the choices in the proof given above to be made without recourse to the Axiom of Choice. The same applies to the countable version of Zorn's Lemma, Theorem 2.18, which also does not need any form of the Axiom of Choice to work. However, the proof of the general form of Zorn's Lemma does need the Axiom of Choice, as the elements of the poset X may not be so conveniently listed as those of a countable set are.

Most published proofs of Zorn's Lemma are quite short but require extra background knowledge in set theory. Here is a proof of Zorn's Lemma from the Axiom of Choice with the minimum of background knowledge required.

Theorem 2.36 (Zorn) *The Axiom of Choice implies Zorn's Lemma.*

Proof Let X be a poset with the Zorn property and for which there is no maximal element. This proof will construct a chain C_0 in X with no upper bound, which obviously contradicts the Zorn property.

We first apply the Axiom of Choice. Considering X as a non-empty set, and P_0 the set of all non-empty subsets of X, by the Axiom of Choice there is a function $f: P_0 \to X$ such that $f(A) \in A$ for all $A \in P_0$. Now let $C \subseteq X$ be a chain. By the Zorn property there is some upper bound, $y \in X$, for C. In other words, the set $U_C = \{y \in X : \forall x \in C\, x \leqslant y\}$ is non-empty and hence in P_0. Thus $f(U_C)$ is an upper bound for C. Composing functions $C \mapsto U_C \mapsto f(U_C)$ we obtain a function u such that $u(C)$ is an upper bound of C whenever $C \subseteq X$ is a chain.

Now we construct our impossible chain C_0 of X. This chain (and others that we consider in the argument) will have the special property that it is *well-ordered*, which means, that it is linearly ordered by the order $<$ on X and that *every non-empty subset of it has a least element*.

Let D be the set of chains $C \subseteq X$ which are well-ordered and for which we have the following holding for every $x \in C$:

$$x = u(\{y \in C : y < x\}).$$

In other words, every element of C should be determined via the function u by its predecessors in C. Note that the empty chain $\varnothing$ is such a chain, so D is not empty. The chain $\{u(\varnothing)\}$ consisting of a single element is also in D by the same reasons.

There are two important facts about D that we must prove.

The first is that if $C \in D$ then the chain $C \cup \{u(C)\}$ formed by adding the canonical choice for an upper bound of C is also in D. Checking the conditions for $C \cup \{u(C)\}$ is quite straightforward. The most tricky one is the well-ordering property; but if $A \subseteq C \cup \{u(C)\}$ is non-empty then either $A \cap C$ is non-empty and has a least element (since C is well-ordered) or else $A = \{u(C)\}$.

The second fact about D is that for any two chains C_1 and C_2 of D we have that one is an initial segment of the other. Here is where we use the well-ordering property. If either C_1 or C_2 is empty there is nothing to prove so assume otherwise. Then the least element of C_1 and the least element of C_2 must both be $u(\varnothing)$, so C_1 and C_2 agree on their least element.

Suppose to start with that there is $x \in C_1$ which is not in C_2. Then there is a least such $x_1 \in C_1 \setminus C_2$, and $C_2 \subseteq \{y \in C_1 : y < x_1\}$. Now assume that there is also a $y \in C_2 \setminus C_1$. Again, take the least such $y_2 \in C_2 \setminus C_1$, and observe that for this y_2 we have $\{z \in C_1 : z < y_2\} = \{z \in C_2 : z < y_2\}$. But $y_2 = u(\{z \in C_2 : z < y_2\})$. There is also a least $y_1 \in C_1$ greater than all elements of $\{z \in C_1 : z < y_2\}$. Clearly $y_1 \neq y_2$ but for this y_1 we have

$$y_1 = u(\{z \in C_1 : z < y_1\}) = u(\{z \in C_2 : z < y_2\}) = y_2$$

which is impossible. So this argument shows that there is in fact no element $y \in C_2 \setminus C_1$, and hence that if there is $x \in C_1$ which is not in C_2 then C_2 is an initial segment of C_1. If there is $x \in C_2$ which is not in C_1 then a similar argument shows C_1 is an initial segment of C_2, and if neither of these applies then $C_1 = C_2$.

These technical properties of D now complete the proof, for the fact that of any two chains in D one is always an initial segment of the other shows that $\bigcup D = \{x \in X : \text{there exists } C \in D \text{ such that } x \in C\}$ is actually a chain. It is also well-ordered, since if $A \subseteq \bigcup D$ is non-empty there is $x \in C \in D$ with $x \in A$ and the least element of A can now be found in $A \cap C$. Therefore $\bigcup D \in D$. But this quickly gives us a contradiction as $\bigcup D \cup \{u(\bigcup D)\}$ is also a well-ordered chain with all the required properties to be in D, but cannot be in D since $u(\bigcup D)$ is greater than all elements of $\bigcup D$. $\qquad\square$

The following direction is much easier.

Theorem 2.37 *Zorn's Lemma implies the Axiom of Choice.*

Proof Given X, a set of non-empty sets, consider the set C of *partial choice functions*, $f : Y \to \bigcup X$ such that $f(x) \in x$ for all $x \in Y$ where $Y \subseteq X$. C is made into a poset by $f < g$ if g extends f. It is straightforward to check that the Zorn property holds and that a maximal element is a required choice function. $\qquad\square$

The Axiom of Choice and a related principle, the Well-Ordering Principle, were around before Zorn, but Zorn's contribution seems to be to provide a useful and strong principle which is equivalent to these that can easily be used in algebra and other settings, without the tricky set theoretical terminology that was then common. From a more practical point of view, the Axiom of Choice is the easiest to understand and justify as an axiom, but as we have seen it can be tricky to use, and the more convenient Zorn's Lemma is usually preferred.

It may be interesting to learn that Zorn's Lemma directly implies König's Lemma. I am not sure how edifying this particular argument is, although it does apply in the most general case discussed in the previous chapter, and it does provide a useful link between the two chapters. We will return to this point in the next chapter and give a less direct but more illuminating argument, culminating with Theorem 3.13, showing that Zorn's Lemma implies König's Lemma.

Theorem 2.38 *Any infinite finitely branching tree has an infinite path.*

Proof (Sketch) We consider the set X of all *infinite subtrees* S of T. This is non-empty as it contains T itself. For the ordering we take, rather unusually, the reverse of $\subseteq$, that is we define $S_1 \leqslant S_2$ if and only if $S_1 \supseteq S_2$.

You should be able to convince yourself that a $\leqslant$-maximal subtree (i.e. a $\subseteq$-*minimal* subtree) is actually a path. This is like the argument in the previous chapter. If it is not in fact a path and has some branching, then we can find an infinite subtree and hence show the tree is not maximal.

The awkward bit is to show that the poset x has the Zorn property. If $C \subseteq X$ is a chain of infinite subtrees it is fairly easy to show that $Y = \bigcap C$ is also a subtree. In fact Y is also infinite, though this takes a little bit of proving. The trick required is to note that T has only finitely many nodes at each level n, and hence a subtree of T is infinite if and only if it has at least one node at each level n. This applied to $\bigcap C$ since each tree in C has only finitely many nodes at each of the levels, but all levels are represented.   □

For the most general form of König's Lemma, some form of the Axiom of Choice is necessary, but there are weaker forms that suffice to prove König's Lemma (though not, of course, Zorn's Lemma). A full discussion of this will take us too far off track and the reader is directed to set theory texts, such as Lévy's *Basic Set Theory* [8].

3

Formal systems

3.1 Formal systems

Formal systems are kinds of mathematical games with strings of symbols and precise rules. They mimic the idea of a 'proof'. This chapter introduces formal systems through an example that turns out to be closely connected with König's Lemma. This simple example is based on the trees that we studied earlier. Formal systems are the 'arguments from limited knowledge' that we talked about earlier, and working in them is like being the ant following a tree who cannot see beyond the immediate node it happens to be at.

The particular system that we shall look at here will put some more detail on the ideas introduced earlier about 'two ways of doing it' and how they can be played off against each other to advantage. It is based on finite sequences, or *strings*, of 0s and 1s. The set of all such strings is denoted 2^* or $2^{<\omega}$ and, as we have seen, this set can be regarded as a full binary tree. We shall write the *empty string* of length zero as $\perp$.

Now consider a game starting from a subset $\Sigma \subseteq 2^*$ with the following rules specifying when a string may be written down.

- (Given Strings Rule) You may write down any string σ in Σ.
- (Lengthening Rule) Once a string σ has been written down, you may also write down one or both of the strings $\sigma 0$ or $\sigma 1$.
- (Shortening Rule) For any string σ, once you have written down both $\sigma 0$ and $\sigma 1$ then you may write down σ.

These rules may be applied finitely many times, in any order, to any string σ. The *objective* of the game is a further string $\tau \in 2^*$. We want to know, given Σ and τ, whether it is possible to write down τ following the rules above.

Definition 3.1 Let $\Sigma \subseteq 2^*$ and $\tau \in 2^*$. We write $\Sigma \vdash \tau$ to mean that it is possible

24

to write down τ in a finite number of steps that follow the rules of the game for Σ.

If $\Sigma \vdash \tau$ then there is a list of strings that can be written down in the game, each of which is written down according to one of the three rules, the last one in the list being τ. Sometimes this list of strings is called a formal proof or formal derivation of τ from strings in Σ following the rules given. Thus $\Sigma \vdash \tau$ can be expressed as saying 'there is a formal proof of τ from strings in Σ'.

Having got the rules of the game, we might see what we can do with it. For example you should be able to see that $\{00, 01\} \vdash 0$ and $\{00, 01\} \vdash 00100$, and you might be able to convince yourself that in fact from $\{00, 01\}$ we can write down any string starting with 0.

Is it possible to write down 1 starting from $\{00, 01\}$? It is easy to guess the answer should be 'no', but to say $\{00, 01\} \not\vdash 1$ is a precise mathematical statement, and one that should be proved carefully. It is not sufficient to write down some formal derivation from $\{00, 01\}$ and note that this derivation does not include the string 1, since there are infinitely many formal derivations to be considered and $\{00, 01\} \not\vdash 1$ says that *none* of them contain 1. In general, to prove rigorously a statement like 'there is no formal derivation of τ' is often difficult, and is almost always achieved using mathematical induction with some cleverly chosen induction hypothesis, if at all. (Later on, we will have the Soundness Theorem that can be useful in such situations to avoid the induction argument. But in fact the Soundness Theorem itself is proved by induction.)

Proposition 3.2 *Suppose that $\tau \in 2^*$ and $\{00, 01\} \vdash \tau$. Then τ must start with 0.*

Proof Consider a formal derivation of τ from $\{00, 01\}$. This derivation has finitely many steps in it. We shall do induction on the number of steps in such a formal derivation.

Our *induction hypothesis* $H(n)$ is that if τ has a formal derivation from $\{00, 01\}$ with *at most n* steps then τ must start with a 0.

To see that $H(1)$ is true observe first that the only formal derivations with one step are those that write down a single $\sigma \in \Sigma$. But all strings in $\{00, 01\}$ start with 0, so all such σ that we could get in one step start with 0, and hence $H(1)$ is true.

Now suppose that $H(n)$ is true and τ has a formal derivation from $\{00, 01\}$ with $n + 1$ steps. If the last step is 'we write down τ because $\tau \in \{00, 01\}$' things are easy as all these strings start with 0. If the last step is 'we write down τ because $\tau = \sigma i$ and we have already written down σ' (where i is 0 or

1) then we can use our induction hypothesis. Since σ has already been written down, it has a derivation of at most n steps so by $H(n)$ σ starts with a 0, and hence $\tau = \sigma i$ also starts with 0. Finally, if the last step is 'we write down τ because we have already written down $\tau 0$ and $\tau 1$' then we use $H(n)$ again. Since both $\tau 0$ and $\tau 1$ have derivations of length at most n they both must start with 0. In particular $\tau 1$ must start with 0. That means that τ is a non-empty string that starts with a 0, as required. This proves $H(n+1)$ and completes our proof by induction. $\qquad\square$

It follows that $\{00, 01\} \nvdash 1$ as 1 does not start with 0, and this finally answers our question.

Quite a lot more can be said about formal proofs in this system. To handle infinite sets Σ we have the following proposition.

Proposition 3.3 *Suppose* $\Sigma \subseteq 2^*$ *and* $\Sigma \vdash \tau$. *Then there is a finite subset* $\Sigma_0 \subseteq \Sigma$ *such that* $\Sigma_0 \vdash \tau$.

Proof A formal derivation is a finite list of strings, so the Given Strings Rule can only be used finitely many times. Let Σ_0 be the set of strings from Σ for which the Given Strings Rule is used in a derivation of τ. Then exactly the same formal derivation shows $\Sigma_0 \vdash \tau$. $\qquad\square$

Note that the particular set $\Sigma_0 \subseteq \Sigma$ in the last proposition depends on τ. It is not necessarily the case that there is a finite $\Sigma_0 \subseteq \Sigma$ such that $\Sigma_0 \vdash \tau$ for *all* τ that can be derived from Σ.

Formal derivations or proofs are finite mathematical objects, and as such are the objects of a mathematical theory. This is because we have specified exactly what rules are going to be allowed in a proof, and not left it to the subjective judgment of another human being. Indeed, the branch of mathematical logic called *proof theory* studies proofs as mathematical objects. The following is a somewhat typical result in the proof theory of the particular rather simple formal system being discussed here.

Proposition 3.4 *Suppose* $\Sigma \subseteq 2^*$ *is finite and* $\Sigma \vdash \tau$. *Then there is a derivation of* τ *from* Σ *taking the following form.*

(i) *First strings from* Σ *are written down, using the Given Strings Rule.*

(ii) *Next any required applications of the Shortening Rule are made.*

(iii) *Finally none, one or more applications of the Lengthening Rule are used as necessary to derive* τ.

Proof By induction on the number of steps in a formal derivation. We start by assuming that we have a formal derivation of τ from Σ of length $n+1$ and inductively assume that whenever σ has a derivation from Σ of length n then there is a derivation of σ from Σ in the form required. We now transform the formal derivation of τ from Σ using this induction hypothesis.

The base case of the induction is when the derivation has length 1. In this case, the statement proved is some $\sigma \in \Sigma$ using the Given Strings Rule, and this proof is of the required form, as the Lengthening and Shortening Rules need not be used at all for it to be in the correct form. More generally, any proof of τ in which the last step is the Given Strings Rule can be rewritten as a one-line proof of τ using the Given Strings Rule only.

Suppose τ is derived in length $n+1$ where the last step is a derivation of τ from some statement ρ by the Lengthening Rule. By induction, there is a proof of ρ of the required form, where ρ is derived in the last step. Then we get a derivation of τ by appending to this proof a single step using the Lengthening Rule.

The only tricky case is when the last step is the Shortening Rule. Here τ is derived from previous statements $\tau 0$ and $\tau 1$ by this rule. By our induction hypothesis there is a proof from Σ of $\tau 0$ in three blocks, B_1, B_2, B_3, where B_1 uses Given Strings only, B_2 uses Shortening only, and B_3 uses Lengthening only, and B_2, B_3 may be empty. Similarly there is a proof from Σ of $\tau 1$ in three blocks, B_1', B_2', B_3'. If the last step in either of these proofs of $\tau 0$, $\tau 1$ is the Lengthening Rule, then the statement τi would have been deduced from τ itself, so there would be a proof of τ of the required form obtained by removing this last step. Otherwise, the Lengthening Rule is never used in these proofs and blocks B_3 and B_3' are absent. In this case a valid proof of τ in the required form can be obtained by writing down B_1, B_1', B_2, B_2' (which derive both $\tau 0$ and $\tau 1$) and then following this by a single application of the Shortening Rule to derive τ. $\qquad\square$

Formal systems typically have a mixture of rules, some of which make the strings shorter and some of which make the strings longer. The particular system here is rather straightforward in that any formal derivation can be rewritten to use rules that shorten strings first, and then all rules to lengthen strings last. This makes it rather easier to see what can and cannot be derived from a given set Σ. In particular, the last proposition can be used to answer any question of the form, 'does $\Sigma \vdash \tau$?' for finite sets Σ. Most other systems are rather harder to work with than this one.

Here is a nice application of this result that, under certain conditions, allows us to eliminate an unnecessary assumption in a formal derivation.

Proposition 3.5 *Suppose $\Sigma \subseteq 2^*$, $\sigma, \tau \in 2^*$ and both $\Sigma \cup \{\tau 0\} \vdash \sigma$ and $\Sigma \cup \{\tau 1\} \vdash \sigma$. Then $\Sigma \vdash \sigma$.*

Proof We suppose that $\Sigma \nvdash \sigma$ and by looking at formal derivations of σ from $\Sigma \cup \{\tau i\}$ ($i = 0, 1$) we shall show that both $\tau 0$ and $\tau 1$ are initial segments of σ. This is of course impossible!

To this end, consider a formal derivation of σ from $\Sigma \cup \{\tau 0\}$ in which all applications of the Given Strings Rule come first, then all applications of the Shortening Rule are used to derive an initial part of σ, and finally the Lengthening Rule is applied to obtain σ. Since $\Sigma \nvdash \sigma$ the string $\tau 0$ must be used somewhere. Consider in particular the *first* place where it is used. If this is as an assumption to the Shortening Rule, then $\Sigma \vdash \tau 1$ since $\tau 1$ is required as the other assumption in the Shortening Rule and $\tau 0$ was not used earlier. This would mean that $\Sigma \vdash \sigma$ as $\Sigma \cup \{\tau 1\} \vdash \sigma$. Therefore $\tau 0$ is not in fact used in the Shortening Rule and hence is used in the Lengthening Rule, so $\tau 0$ is an initial part of σ. The same argument with 0 and 1 swapped shows $\tau 1$ is also an initial part of σ, and this gives our contradiction.   □

Proposition 3.4 has two nice consequences concerning the empty string.

Proposition 3.6 *Suppose $\Sigma \subseteq 2^*$ and $\Sigma \vdash \bot$. Then there is a formal derivation of $\bot$ from Σ using only the Given Strings and Shortening Rules.*

Proof Let $\Sigma_0 \subseteq \Sigma$ be finite such that $\Sigma_0 \vdash \bot$. Then there is a formal proof of $\bot$ from Σ_0 in which any applications of the Lengthening Rule are at the end and are used to derive $\bot$. But the Lengthening Rule increases the length of a string and $\bot$ has the least possible length (zero) so there are no applications of the Lengthening Rule in such a formal derivation.   □

Proposition 3.7 *Suppose $\Sigma \subseteq 2^*$ and $\Sigma \vdash \bot$. Then there is a finite subset $\Sigma_0 \subseteq \Sigma$ and $n \in \mathbb{N}$ such that for all strings $\sigma \in 2^*$ of length n, $\Sigma_0 \vdash \sigma$ using only the Given Strings and Lengthening Rules.*

Proof Let $\Sigma_0 \subseteq \Sigma$ be finite and such that $\Sigma_0 \vdash \bot$ using only the Given Strings and Shortening Rules. Let Σ_1 be the set of all strings derivable from Σ_0 using only the Given Strings and Shortening Rules. Let n be the maximum length of strings in Σ_0 and let σ have length n. Note that each string in Σ_1 has length at most n as the Shortening Rule shortens strings. We claim that σ can be derived from some $\tau \in \Sigma_0$ using the Lengthening Rule only.

To see this, consider the string $\tau \in \Sigma_1$ of greatest length that is equal to σ

or to an initial segment of σ. There must be such a τ as $\perp \in \Sigma_1$. If τ is in Σ_0 we are done; otherwise τ was derived from strings $\tau 0$ and $\tau 1$ in Σ_1 by the Shortening Rule, and the length of τ is strictly less than n as the Shortening Rule is a string-shortening rule. But then one of $\tau 0$ or $\tau 1$ must be an initial segment of σ, since τ is already an initial segment of length less than n, the length of σ. This contradicts the maximality of the length of τ. $\qquad \Box$

We have now explored this formal system in some detail. Most of our results have been mathematical theorems and arguments concerning what does or does not have a *formal proof*. Some people like to call such mathematical theorems and arguments *metatheorems* and *metaproofs* since they are theorems and proofs *about* theorems and proofs. This terminology is sometimes useful and you may use it if you wish, but there is nothing special about such 'metamathematics'. Metamathematics is just the ordinary mathematics you are already used to.

To take this study further, and in particular to apply it to other areas of mathematics, we have to start thinking about what the formal system actually *means*. A formal system is really just like a game with symbols. To prove anything in this system you just follow the rules, nothing more. This is in some ways good, because there is no thinking involved and because a machine could check your working for you, but in other ways it is bad, because it does not seem that this system is proving anything particularly useful, or *about* anything. So the next step is to try to attach meaning or *semantics* to the symbols we are using, and interpret the strings of symbols of the system.

It may not be clear that there is anything useful to say about this particular system. After all, what could a rather boring string of zeros and ones really mean? (We will look at a particular non-obvious but rather nice interpretation in a moment.) A more subtle point, and one that is potentially very fruitful to consider, is that we might be able to find *more than one* possible set of meanings to the system. Just because we may choose to focus on one particular meaning to the symbols does not mean that there are no other equally valid semantics that are possible or useful. In fact, the particular system we have here will later on be shown to be a special case of the system of propositional logic involving 'not', 'and' and 'or' of Chapter 6.

If we are to attach any semantics at all, we need to interpret $\Sigma \vdash \sigma$ in some other mathematical way. We shall use the notation $\Sigma \vDash \sigma$ for this interpretation and then prove as part of our theory *about* this formal system that the two notions $\vdash$ and $\vDash$ are equivalent. I want to relate the formal system here with trees and infinite paths in a tree, so I propose the following somewhat complicated looking definition of semantics.

Definition 3.8 Let $\Sigma \subseteq 2^*$ and $\sigma \in 2^*$. Write $\Sigma \vDash \sigma$ for the statement

- whenever $p \subseteq 2^*$ is an infinite path which passes through σ then p passes through some $\tau \in \Sigma$.

Be careful reading this statement, especially looking out for the 'whenever', and the implicit 'for all infinite paths p' in it.

Example 3.9 If $\Sigma = 2^*$ and $\sigma \in 2^*$ then any infinite path passing through σ passes through some $\tau \in \Sigma$, namely σ itself. So $\Sigma \vDash \sigma$. More generally, this argument shows that $\Sigma \vDash \sigma$ whenever $\sigma \in \Sigma$.

Example 3.10 Suppose σ is the root $\perp$. All infinite paths pass through the root. So $\Sigma \vDash \perp$ holds if and only if for every infinite path there is some $\tau \in \Sigma$ that the path passes through.

The next step is to prove that $\Sigma \vdash \sigma$ and $\Sigma \vDash \sigma$ are equivalent, i.e. each one implies the other. The two directions of implication here are called the *Soundness* and *Completeness Theorems*.

Theorem 3.11 (Soundness) *Let* $\Sigma \subseteq 2^*$ *and* $\sigma \in 2^*$ *and suppose that* $\Sigma \vdash \sigma$. *Then* $\Sigma \vDash \sigma$.

Proof By induction on the length of formal derivations of σ. Our induction hypothesis $H(n)$ is that, whenever $\Sigma \vdash \sigma$ with a derivation of length at most n then $\Sigma \vDash \sigma$. This is true for derivations of length one since $\sigma \in \Sigma$ implies $\Sigma \vDash \sigma$ by Example 3.9 above.

Now suppose $\Sigma \vdash \sigma$ with a formal derivation of length $n+1$ and in which the last step is the Lengthening Rule, so $\sigma = \tau i$ where τ is derived in at most n steps and $i = 0$ or 1. Suppose also $p \subseteq 2^* \setminus \Sigma$ is an infinite path passing through σ. Then p must also pass through τ, since τ has length one less than that of σ and so, by $H(n)$, the path p passes through some element of Σ.

Finally suppose $\Sigma \vdash \sigma$ with a formal derivation of length $n+1$ and in which the last step is the Shortening Rule, so both $\sigma 0$ and $\sigma 1$ are derived with at most n steps. Suppose again that p is an infinite path passing through σ. Then p must pass through one of $\sigma 0$ or $\sigma 1$, so by the induction hypothesis $H(n)$ p must pass through some element of Σ. $\qquad\square$

Theorem 3.12 (Completeness) *Let* $\Sigma \subseteq 2^*$ *and* $\sigma \in 2^*$ *and suppose that* $\Sigma \vDash \sigma$. *Then* $\Sigma \vdash \sigma$.

Proof We start by assuming that $\Sigma \nvdash \sigma$ and find an infinite path $p \subseteq 2^* \setminus \Sigma$ that contains σ. This will show that $\Sigma \nvdash \sigma$, and hence give the contrapositive of the required theorem.

Let X be the set of all supersets $\Sigma' \supseteq \Sigma$ such that $\Sigma' \nvdash \sigma$, and order X by normal inclusion, $\subseteq$. This makes X into a poset, and it has the Zorn property. This is because if $C \subseteq X$ is a chain then firstly $\bigcup C \supseteq \Sigma$, since each element of C contains Σ as a subset, and secondly if $\bigcup C \vdash \sigma$ then there is a formal derivation of σ from some finite subset $\Sigma_0 \subseteq \bigcup C$; but each element σ_i of Σ_0 is in some set C_i in C and there are only finitely many such σ_i so all σ_i are in some $\Sigma_j \in C$, since C is a chain.

By Zorn's Lemma, X has a maximal element Σ^+. We shall show that Σ^+ is the complement of an infinite path that contains σ. Putting $p = 2^* \setminus \Sigma^+$ will then complete the proof.

To see that $\sigma \notin \Sigma^+$ it suffices to observe that $\Sigma^+ \cup \{\sigma\} \vdash \sigma$ by the Given Strings Rule, and so $\Sigma^+ \neq \Sigma^+ \cup \{\sigma\}$, because the latter is not in X.

Note next the following argument using maximality of Σ^+: if τ is any string then $\tau \in \Sigma^+$ if and only if $\Sigma^+ \cup \{\tau\} \nvdash \sigma$. For one direction, if $\tau \in \Sigma^+$ then $\Sigma^+ \cup \{\tau\} = \Sigma^+ \nvdash \sigma$ as $\Sigma^+ \in X$. Conversely, if $\Sigma^+ \cup \{\tau\} \nvdash \sigma$ then $\Sigma^+ \cup \{\tau\} \supseteq \Sigma^+$ is in X so these sets must be equal by the maximality of Σ^+, and hence $\tau \in \Sigma^+$.

To see that $2^* \setminus \Sigma^+$ is a tree, suppose $\tau i \notin \Sigma^+$ where i is 0 or 1. We show that $\tau \notin \Sigma^+$. By assumption and maximality, $\Sigma^+ \cup \{\tau i\} \vdash \sigma$. This means $\Sigma^+ \cup \{\tau\} \vdash \sigma$ by the same derivation and one extra step using the Lengthening Rule, so $\tau \notin \Sigma^+$ as required.

Finally, to see that $2^* \setminus \Sigma^+$ is an infinite path, suppose $\tau \notin \Sigma^+$ and we attempt to show that $\tau i \notin \Sigma^+$ for exactly one i from 0 or 1. By maximality, $\Sigma^+ \cup \{\tau\} \vdash \sigma$. If *both* $\tau 0, \tau 1 \in \Sigma^+$ we would have $\Sigma^+ \vdash \sigma$ by the same derivation and an application of the Shortening Rule. This is impossible as $\Sigma^+ \in X$. Furthermore, if *neither* of $\tau 0, \tau 1$ is in Σ^+ we would have $\Sigma^+ \cup \{\tau i\} \vdash \sigma$ for each $i = 0, 1$ by the same maximality argument, and this implies $\Sigma^+ \vdash \sigma$ by Proposition 3.5. Thus exactly one of $\tau 0, \tau 1$ is in Σ^+ and τ has a unique extension τi not in Σ^+. $\qquad \square$

Theorem 3.13 (König's Lemma) *An infinite tree $T \subseteq 2^*$ contains an infinite path.*

Proof Let $T \subseteq 2^*$ be a tree and let $\Sigma = 2^* \setminus T$ be its complement. Since T is a tree and in particular closed 'downwards' by restriction, Σ is closed 'upwards' under applications of the Lengthening Rule.

Now suppose also that T contains no infinite paths. This means that $\Sigma \vDash \bot$ since the complement of Σ does not contain an infinite path. By the Completeness Theorem we have $\Sigma \vdash \bot$. Then by Proposition 3.7 there is a finite subset

$\Sigma_0 \subseteq \Sigma$ and $n \in \mathbb{N}$ such that all strings $\sigma \in 2^*$ of length n are derivable from Σ_0 by the Given Strings and the Lengthening Rules only. But $\Sigma_0 \subseteq \Sigma$ and Σ is closed under the Lengthening Rule, so Σ contains all strings of length n. That means that T does not contain any strings of length n, and as it is closed downwards all strings in T must have length less than n. Hence T is finite. $\square$

This has been a long and detailed discussion of one particular formal system. It was nice that the system has some elegant mathematics and that the now familiar König's Lemma comes for free from the proof-theoretic analysis and the Completeness Theorem. But it may not be obvious at this stage what the formal system has to do with logic, and where the strange idea of semantics we have discussed comes from.

As hinted already, there are many possible answers to this question. The simplest one arises by trying to analyse a problem into many cases. Consider a problem in which all possible situations are described by infinitely many variables $p_0, p_1, \ldots$. Each p_i can take one of two values, 0 and 1 (which you can think of as 'false' and 'true' respectively, if you like). A string $\sigma = s_0 s_1 \ldots s_{k-1} \in 2^*$ represents a situation where $p_i = s_i$ for each $i < k$ and some unknown values of p_i for $i \geqslant k$. Think of this situation as being 'impossible' for some specific reason to do with the problem in hand; the situation is known to be impossible because of the values of p_i for $i < k$, irrespective of the values of p_i for $i \geqslant k$. Then a set Σ of strings represents a set of situations all known to be impossible. Now we may review the proof rules with this interpretation: if $\sigma = s_0 s_1 \ldots s_{k-1} \in 2^*$ is impossible irrespective of what s_k is, then $\sigma 0$ and $\sigma 1$ both represent impossible situations. This is a reasonable justification for the Lengthening Rule. Also, if $\sigma = s_0 s_1 \ldots s_{k-1} 0$ and $\sigma = s_0 s_1 \ldots s_{k-1} 1$ are both impossible, then $\sigma = s_0 s_1 \ldots s_{k-1}$ is also impossible because p_k can only be either 0 or 1, not anything else. And this is a reasonable justification for the Shortening Rule.

Now consider a set Σ listing all impossible situations for the problem at hand. Some situation is possible if it is described by a unique value from 0, 1 for every p_i, and no finite string of such values is in Σ. Thus a situation is possible if it is described by an infinite path that avoids Σ completely. This is clearly related to the notion of semantics we were studying above. In particular, $\Sigma \vDash \sigma$ means that if every $\tau \in \Sigma$ is impossible then so is σ. The special case of this, $\Sigma \vDash \perp$, means that every situation is impossible, since the empty string does not specify any value for any p_i. We will use the symbol $\perp$ throughout the rest of the book to denote the impossible situation, i.e. to denote a contradiction.

As already mentioned, the system here is a fragment of 'ordinary logic'; see Example 7.27 where it is explained how to embed this system into propositional logic (a traditional system of logic involving the connecting words 'not', 'and' and 'or'). In fact slightly more is true: we could formulate the whole of propositional logic in the way indicated in this chapter using trees and 'possible situations'. Example 7.28 gives some details. But correct as this might be, it is not the whole story, as logic should also relate to natural mathematical deductions in systems looking very similar to the actual kinds of arguments used as part of mathematical work. The system of this chapter and this retrospective overview of it seems to show that, at a minimum, some sort of case-by-case analysis is required for correct deductions, and a tree is a useful tool for analysing such deductions. In the next chapter we shall look at a more natural alternative technique for proof, that replaces the mode of proof based on a tree of 'bad' or 'impossible' situations with the very familiar rule of 'Reductio Ad Absurdum' instead.

3.2 Examples and exercises

The main point of this chapter has been to present an example of a formal system and show that the analysis of even quite simple formal systems can be mathematically interesting. For beginners, perhaps the most important thing to learn is the way induction on length of derivations is used. This is the focus of the first exercises.

For the next few exercises, σ ranges over elements of 2^* and Σ_0 is the set

$$\{0, 10, 110, 1110, \ldots\}.$$

Exercise 3.14 Prove that $\Sigma_0 \vdash \sigma$ implies that σ contains at least one 0. (Use induction on the length of a derivation.)

Exercise 3.15 Show that $\Sigma_0 \nvdash \bot$.

Exercise 3.16 Let $\sigma = 111\ldots1$ be a string of 1s (i.e. not containing any 0). Use induction on the length of σ to show that $\Sigma_0 \cup \{\sigma\} \vdash \bot$.

Exercise 3.17 Let σ be a string containing at least one 0. Show that $\Sigma_0 \vdash \sigma$.

Exercise 3.18 Prove that there is a unique maximal element Σ^+ in

$$X = \{\Sigma \subseteq 2^* : \Sigma_0 \subseteq \Sigma \text{ and } \Sigma \nvdash \bot\}.$$

(Here, X is ordered by $\subseteq$.) Say what this maximal element Σ^+ is. Say why it is maximal and the only such maximal element in X. (Hint: Zorn's Lemma will not help you here. Argue directly.)

Exercise 3.19 Say what the path $p = 2^* \setminus \Sigma^+$ is, and justify your assertion.

Exercise 3.20 Use the Soundness Theorem to show that, for all $\sigma \in \Sigma_0$,

$$\Sigma_0 \setminus \{\sigma\} \nvdash \sigma.$$

(Hint: define a suitable infinite path passing through σ.)

Thus Σ_0 is an infinite non-trivial set of strings which does not prove the whole of 2^* and for which each $\sigma \in \Sigma_0$ cannot be removed without changing the set of consequences of Σ_0. The fact there are such sets is quite interesting in itself. For other systems, the existence of such sets is often a particularly challenging question.

Exercise 3.21 Say that a set $\Sigma \subseteq 2^*$ is *consistent* if $\Sigma \nvdash \bot$. Give as many possible interpretations as you can of what it means to say that Σ is consistent. If you use the Completeness Theorem or the Soundness Theorem anywhere, state which one you use at the point where it is needed.

Exercise 3.22 Give an algorithm that decides, on input a finite set $\Sigma \subseteq 2^*$ and $\sigma \in \Sigma^*$, whether $\Sigma \vdash \sigma$. Prove your algorithm always gives the correct answer.

The next exercises discuss other formal systems, and the fun of investigating these is left up to you.

Exercise 3.23 A system P has symbols $|$, $-$ and $=$ and the following rules.

- You may write down $|-|=||$.
- If you have written down σ you may write down $|\sigma|$.
- If you have written down $\sigma = \tau$ you may write down $\sigma | = | \tau$.

Suggest a possible semantics for this system and state and prove the corresponding Soundness and Completeness Theorems.

Exercise 3.24 In the MIU system (Hofstadter [6]) there are three symbols M, I, and U. These can be combined to form strings of symbols, such as MIUUII. The rules are as follows.

- You may write down MI.
- If you have written down σI you may write down σIU.
- If you have written down Mσ then you may write down M$\sigma\sigma$.

- If you have written down $\sigma\text{III}\tau$ then you may write down $\sigma\text{U}\tau$.
- If you have written down $\sigma\text{UU}\tau$ then you may write down $\sigma\tau$.

Which of the following can be derived using these rules?

(i) MIU
(ii) MUIMUI
(iii) MIUU
(iv) MUIUIU
(v) MIUUIIUUI
(vi) MIIIIIIII
(vii) MIIIIIII
(viii) MIIIIII
(ix) MU

Justify all your assertions by providing proofs.

Exercise 3.25 'But that's semantics. I'm not here to discuss semantics!' How many times have you heard a politician (or anyone else) say that? Assuming that politicians do not concern themselves with semantics at all, discuss what it is that politicians actually do.

3.3 Post systems and computability*

There have been many attempts to define the notion of 'formal system' and study them in generality, rather than on a case-by-case basis. Unfortunately the notion of 'formal system' is a little difficult to pin down. For example, it is natural to focus on systems based on strings of symbols taken from a finite alphabet, though some of the examples towards the end of this book will require us to look at very large infinite alphabets too.

With this proviso in mind, we can present here a large family of formal systems, called Post systems after Emil Post, which cover a surprisingly large range of applications.

Definition 3.26 A *Post system* comprises a finite set of symbols A, called the *alphabet*, another finite set of symbols V, disjoint from A and called the set of *variables*, and a finite set of rules of the form

$$\frac{\sigma_1, \ldots, \sigma_k}{\tau}$$

where $\sigma_1, \ldots, \sigma_k, \tau$ are finite strings of symbols from $A \cup V$ and $k \in \mathbb{N}$ is possibly 0. The idea is that this rule represents the derivation rule 'if $\sigma_1, \ldots, \sigma_k$ have been written down (where each variable in the σ_i represents an arbitrary finite string of symbols from A), then τ may be written down (with any variables in τ substituted by the appropriate string from A)'. The strings σ_i are called the *premises* of the rule and τ is the *conclusion*.

These rules replace the Lengthening and Shortening Rules of the main system discussed in this chapter. In fact, Lengthening and Shortening are special cases, and can be written as

$$\frac{x}{x0}, \frac{x}{x1}, \frac{y0, y1}{y}$$

in the new notation, where x, y are variables. Note that although these variables represent arbitrary strings of 0s and 1s, each instance of a variable must represent the same string in each usage of the rule.

The Given Strings Rule, which says that in a derivation from a set Σ you may write down any $\sigma \in \Sigma$, is available and present in all Post systems.

A great number of systems, even ones that are not themselves Post systems, can be simulated by Post systems. The usual trick is to take a system S with a finite alphabet, and add new symbols to the alphabet. Then by careful addition of further symbols and rewriting the rules of S by rules in the Post style, it is often possible to come up with a Post system S_P such that $\Sigma \vdash \sigma$ in S holds if and only if $\Sigma \vdash \sigma$ in S_P. The following example gives the basic idea, but more complex examples require some skill to find the right alphabet and proof rules.

Example 3.27 Developing on the system of Exercise 3.23, we devise a formal system with symbols $|$, $*$ and $=$ and the following rules.

- You may write down $|*|=|$.
- If you have written down $\sigma * \tau = \rho$ you may write down $\sigma | * \tau = \chi$, provided $\tau - \rho = \chi$ is also provable in the system of Exercise 3.23.
- If you have written down $\sigma * \tau = \rho$ you may write down $\sigma * \tau | = \chi$, provided $\sigma - \rho = \chi$ is also provable in the system of Exercise 3.23.

This is not itself a Post system, but can be simulated by a Post system. The trick is to add the extra symbol $-$ to the alphabet and combine the rules of the two systems into one large system. In this case the Post-style rules are

$$\frac{}{|-|=||}, \frac{x}{|x|}, \frac{x=y}{x|=|y},$$

and

$$\frac{}{|*|=|}, \frac{x*y=z, \ y-z=w}{x|*y=w}, \frac{x*y=z, \ x-z=w}{x*y|=w}$$

where x, y, z, w are variables. (Note the use of Post rules with empty premises on the top to get us started. This is allowed in the definition.)

Post systems, like the vast majority of other formal systems on finite alphabets, are partially *computable*. A computer program can be written to check that a formal derivation does indeed follow all the rules precisely, and also, given a finite input set Σ it is possible to write a computer program that generates all possible σ such that $\Sigma \vdash \sigma$. Such a program starts by writing down each string in Σ and then repeatedly checks every combination of strings written down and every rule to see whether that rule can be applied, and if so writes down the result. Thus the computer program will generate a (possibly endless) sequence of strings.

This gives a *partial* answer to Exercise 3.22 on decidability in the case of a general Post system: to see whether $\Sigma \vdash \sigma$ we start our program and wait to see whether σ is written down. If σ does have a formal derivation it will indeed be written down and we will have our answer. However, if σ does not have a formal derivation our computer program most likely will continue for ever. Thus the program will never tell us that σ is not ever written down. So this program does not provide a full algorithm deciding provability in the Post system. In fact there are Post systems for which provability is *not* decidable by a computer program.

Post systems provide an elegant introduction to the theory of computability. It turns out that any computer program can also be simulated by a Post system, and thus Post systems are effectively a formal mathematical model of a computer. There are therefore many links between formal systems and computability, and this is a subject that could take up a separate book. Instead of getting side-tracked even further, I direct the reader to several excellent texts that provide more information on computability and its links with logic, for example those by Boolos and Jeffrey [2] and by Cutland [4].

4

Deductions in posets

4.1 Proving statements about a poset

The last chapter set out a formal system for which we had to work quite hard to find a mathematically interesting semantics. In this chapter we study another formal system for which the intended semantics is clear: it is about deductions one can make in a poset or family of posets. In fact, the proof system we shall give here provides a formal version of the sorts of arguments you may already have used in an informal way when formulating an argument about a poset.

Amongst other things, we introduce here the important idea of *subproofs* and the *Reductio Ad Absurdum Rule* that will be used in this system and other systems throughout the rest of this book. Our logical analysis of posets will result in some useful, though technical, existence theorems. As a corollary we will deduce a nice result about converting a partial order into a linear order.

Suppose we have a poset X and elements $a, b, c, d \in X$. We might be interested in determining which of these four elements is less than or greater than which of the others. If we know a small number of facts about them, we can sometimes deduce other facts.

For example, if we know that $a < b$ and $b < c$ then we can deduce that $a < c$ by the transitivity axiom of a partial order. Similarly, if we know that $b < c$ and $d \not< c$ we can deduce that $d \not< b$.

Can we describe *all* such deductions somehow? And more importantly, can we put our theory of deductions to good mathematical use in the theory of partial orders – similar to the way in which König's Lemma was connected to the theory of trees? The answer to these questions turns out to be yes, and such a theory of deductions can be an important tool in understanding some rather complicated partial orders.

We start by formulating some of the methods by which such statements can be proved or deduced. In doing this I shall change the meaning of the symbol $\perp$

and will use it in this chapter to mean 'contradiction' or 'false'. It is possible that you are used to a different symbol to mean 'contradiction' but the use of a contradiction in 'proof by contradiction' or *Reductio Ad Absurdum* should be familiar enough. These proofs all follow the pattern: if from some assumption A it is possible to prove a contradiction, $\bot$, then the assumption A must be false.

We shall now present a list of rules of formal proof about a poset. In doing, we imagine a non-empty set X. This set X may already be a poset, i.e. have an order relation $<$ on it, or we might be asking hypothetical questions such as 'what happens if $b < c$ and $d \not< c$?' Therefore it helps to have two separate symbols for the order, and we use $\prec$ for the symbol in our formal system, and $<$ for a real partial order on X (if there is any). Normally, we will be given a set of statements or assumptions about the relationships between certain elements of X. These 'given' statements may be all of the possible statements true of the poset X, or only some of them, or they may include statements that are not in fact true at all, but we wish to test to see what their consequences might be.

Definition 4.1 Let X be a poset, or just a non-empty set. We consider strings of the form $a \prec b$ and of the form $a \not\prec b$, where a, b range over elements of X and where $\prec$ and $\not\prec$ are distinct symbols. The rules of proof about the poset X are as follows.

- (Given Statements Rule) If a statement is in our set of given statements or assumptions then it may be deduced (i.e. written down) directly.
- (Transitivity Rule) From statements $a \prec b$ and $b \prec c$ we may deduce $a \prec c$.
- (Irreflexivity Rule) From $a \prec a$ we may deduce $\bot$.
- (Contradiction Rule) From $a \prec b$ and $a \not\prec b$ we may deduce $\bot$.
- (Reductio Ad Absurdum (RAA) Rule) If it is possible to deduce $\bot$ from the current assumptions together with an additional assumption s (of the form $a \prec b$ or $a \not\prec b$) then the opposite, or *negation*, of s (i.e. $a \not\prec b$ or $a \prec b$, respectively) may be deduced *without* the additional assumption.

Note particularly that any assumption whatsoever may be introduced as the new assumption of a Reductio Ad Absurdum Rule. It is usual, however, to introduce the negation of the statement that you are trying to prove.

Formal proofs are, just as in the last chapter and elsewhere in this book, finite. That is, if some statement can be deduced from a set of given assumptions then there is a finite formal proof of this statement that consists of a list of other statements taking part in the deduction. But a few remarks about the

Reductio Ad Absurdum Rule are necessary here. When it is used in a formal proof it is necessary to indicate a *subproof* showing the new assumption and deductions from it. So proofs in a system with Reductio Ad Absurdum are not simple lists of statements, each following from previous statements, but are structured lists of statements and subproofs. Note too that the Reductio Ad Absurdum Rule, when applied carefully, can be *nested*, so a subproof can contain other subproofs. In other words the deduction of ⊥ from a statement *s* may already contain another 'subproof' by Reductio Ad Absurdum from some other assumption *t*. If you examine textbooks (including this one) you will find many such double proofs by contradiction, written in an informal style. When writing proofs formally, I will denote subproofs and assumptions with a vertical line indicating the part of the proof where the assumption is valid.

We will look at some examples of proofs next. The first is quite straightforward as it does not contain any subproofs.

Example 4.2 Let $X \supseteq \{a, b, c\}$ and consider the following statements as assumptions: $a \prec b$, $b \prec c$, and $c \prec a$. Then these assumptions can prove ⊥.

Formal proof

$a \prec b$	(1)	Given
$b \prec c$	(2)	Given
$a \prec c$	(3)	Transitivity
$c \prec a$	(4)	Given
$a \prec a$	(5)	Transitivity
⊥	(6)	Irreflexivity

A simple modification of the above example gives the following.

Example 4.3 Let $X \supseteq \{a, b, c\}$. Then from assumptions $a \prec b$ and $b \prec c$ we may deduce $c \nprec a$.

Formal proof

$a \prec b$	(1)	Given
$b \prec c$	(2)	Given
$a \prec c$	(3)	Transitivity
$\quad c \prec a$	(4)	Assumption
$\quad a \prec a$	(5)	Transitivity
$\quad ⊥$	(6)	Irreflexivity
$c \nprec a$	(7)	Reductio Ad Absurdum

Similarly, we have the following.

Example 4.4 Let $X \supseteq \{a, b\}$. Then from the assumption $a \prec b$ we can prove $b \not\prec a$.

Formal proof

$a \prec b$	(1)	Given
$\quad b \prec a$	(2)	Assumption
$\quad a \prec a$	(3)	Transitivity
$\quad \perp$	(4)	Irreflexivity
$b \not\prec a$	(5)	Reductio Ad Absurdum

A set of statements Σ involving elements from a set X and a partial order on X can therefore be used to deduce or prove other statements using the above rules. As mentioned already, to 'play the game' of making a proof, there is no requirement that the statements in the assumptions be actually true in the partially ordered set. They are just statements and we are investigating those strings that are available as conclusions in this formal system.

Definition 4.5 Let Σ be a set of statements involving elements from a set X and let σ be $\perp$ or else a single string, $a \prec b$ or $a \not\prec b$ for some $a, b \in X$.

- $\Sigma \vdash \sigma$ means there is a formal proof of the statement σ from the assumptions in Σ and the rules above.
- In particular $\Sigma \vdash \perp$ means that there is a formal proof of the contradiction from Σ. We say Σ is *inconsistent*.
- $\Sigma \not\vdash \sigma$ means there is *no* formal proof of σ from Σ.
- In particular $\Sigma \not\vdash \perp$ means that there is no formal proof of the contradiction from Σ. We say Σ is *consistent*.

Note that, by the rule of given statements above, $\Sigma \vdash \sigma$ for every $\sigma \in \Sigma$. Therefore the set of statements provable from Σ is a superset of the set Σ.

Example 4.6 Let $X \supseteq \{a, b, c\}$. Then $\{c \prec b, a \not\prec b\} \vdash a \not\prec c$.

Formal proof

$c \prec b$	(1)	Given
$a \not\prec b$	(2)	Given
$\quad a \prec c$	(3)	Assumption
$\quad a \prec b$	(4)	Transitivity
$\quad \perp$	(5)	Contradiction
$a \not\prec c$	(6)	Reductio Ad Absurdum

We therefore have a system of formal proof, in fact one separate system for each set of individuals X. We have been able to write down and deduce facts like $a \prec b$ and $a \not\prec b$ from a set of assumptions Σ, irrespective of whether our assumptions are true or not, or whether a really is less than b or not. We shall now attach meanings or semantics to our formal system. To do this we imagine all possible partial orderings $<$ on our set X, and for each such partial order on X we consider the set of *all* statements about elements of X that are actually true.

Definition 4.7 Let X be a poset. Then the set of statements *true in X* is defined to be

$$T_X = \{a \prec b : a, b \in X \text{ and } a < b \text{ in } X\} \cup \{a \not\prec b : a, b \in X \text{ and } a \not< b \text{ in } X\}.$$

We would hope that it should at least be the case that such a set T_X is consistent, i.e. $T_X \not\vdash \bot$. This is in fact the case, but is a little harder to prove than might seem at first sight. You might remember from the last chapter that showing some system does not derive some statement is usually proved by induction using a clever induction hypothesis. That is exactly what happens here, and we will prove a more general 'Soundness Theorem' that will give the consistency of T_X as a corollary.

Definition 4.8 We write $\Sigma \vDash \sigma$ to mean: for every poset X, if X makes every statement in Σ true, X also makes σ true.

Theorem 4.9 (Soundness) $\Sigma \vdash \sigma$ *implies* $\Sigma \vDash \sigma$.

Before I give the proof, I must issue a general warning – one that will be applicable on many other similar occasions – about how to read this statement. In particular, if there are *no* posets making the whole of Σ true, as indeed happens with the inconsistent set $\Sigma = \{a \prec b, b \prec c, c \prec a\}$, then the statement of the Soundness Theorem is still valid. For this particular inconsistent set, and any statement σ (or even the statement $\bot$ which is false in all posets) $\Sigma \vDash \sigma$ is a correct assertion because it says that *any poset* that makes Σ true will make σ true. In other words, it says that if at any time you managed to find a poset making Σ true then the statement σ would also be true for this poset. Of course, the reason this holds is simply because you will always fail to find a suitable poset. If no structure makes the whole of Σ true, we shall often say that $\Sigma \vDash \sigma$ holds *vacuously*.

Proof of the Soundness Theorem By induction on the number of steps in a formal derivation. The induction hypothesis $H(n)$ is that if p is a proof in the system given above with at most n steps and p is a proof of σ from a set of given statements Σ, then σ is true in all posets on X that make all the statements in Σ true.

Assume $H(n)$ and that a proof p of σ from Σ has $n + 1$ steps, and consider the very last step in the proof p.

If it is from the 'given statements' rule then σ is in the set Σ so any poset making Σ true automatically makes σ true. (This easy argument also covers the base case $H(1)$ of the induction.)

If the last step uses the transitivity rule then σ is $a \prec c$ where $a \prec b$ and $b \prec c$ have previously been deduced from Σ in at most n steps. Then by $H(n)$ any partial order $<$ on X making Σ true makes both $a \prec b$ and $b \prec c$ true. Thus any such $<$ has $a < b$ and $b < c$ so, by transitivity of $<$, $a < c$ and hence the partial order makes $a \prec c$ true, as required.

If the last step uses the irreflexivity rule then σ is $\perp$ where $a \prec a$ has previously been deduced from Σ in at most n steps. Then by $H(n)$ any partial order $<$ on X making Σ true makes $a \prec a$ true. But a partial order is always irreflexive so cannot make $a \prec a$ true, hence we conclude that there is no such partial order. So as there is no partial order $<$ on X making Σ true it is vacuously the case that every partial order $<$ on X making Σ true also makes $\perp$ true.

If the last step uses the contradiction rule then σ is $\perp$ where $a \prec b$ and $a \not\prec b$ have been deduced from Σ in at most n steps. Then by $H(n)$ any partial order $<$ on X making Σ true makes both $a \prec b$ and $a \not\prec b$ true. Again, the only way this can happen is if there is no such partial order. Then we conclude that as there is no such partial order making Σ true, it is vacuously the case that any partial order making Σ true makes $\perp$ true.

If the last step uses the Reductio Ad Absurdum Rule, it means that σ has been deduced from Σ by first deducing $\perp$ from $\Sigma \cup \{\sigma'\}$, where σ' is the negation of σ. Thus by $H(n)$ every partial order making $\Sigma \cup \{\sigma'\}$ true also makes $\perp$ true, or, put more naturally, there are no partial orders making $\Sigma \cup \{\sigma'\}$ true. In other words any partial order making Σ true must necessarily make σ' false, and hence must make σ true, as required. □

A formal derivation or proof is like the ant's view of the tree in König's Lemma – it is a sort of view from incomplete information about any given poset, and supports exactly the statement it claims to be true. Like König's Lemma, the interesting part here is the converse direction, which in this case is the statement that if something is true in all posets, then there is a formal derivation of it. Such a theorem is called a *Completeness Theorem* as it says that the rules of derivation are sufficient or *complete* for the intended application. By switching the implication round to the contrapositive, completeness theorems can also be seen as mathematical theorems that say certain mathematical structures exist, in this case, saying that certain order relations $<$ exist. We are going to present our Completeness Theorem in this second form first.

Theorem 4.10 (Completeness) *Suppose Σ is a consistent set of statements about elements of a set X. Then there is a partial order $<$ on X making every statement in Σ true.*

Proof Define a partial order $<$ on X by

$$a < b \text{ if and only if } \Sigma \vdash a \prec b.$$

We need to prove that this is a partial order, and that all statements in Σ are made true by this order.

To see that it is a partial order, we check the axioms. First if $\Sigma \vdash a \prec b$ and $\Sigma \vdash b \prec c$ then $\Sigma \vdash a \prec c$ by the transitivity rule of deduction. So $a < b$ and $b < c$ imply $a < c$. For the irreflexivity axiom, consider $a \in X$. It cannot be that $\Sigma \vdash a \prec a$, for this would imply $\Sigma \vdash \bot$ by the irreflexivity rule of deduction, contradicting the consistency of Σ. Therefore $\Sigma \not\vdash a \prec a$ so by our definition it is not true that $a < a$.

To show that every statement in Σ is true for this order, suppose a statement of the form $a \prec b$ is in Σ. Then $\Sigma \vdash a \prec b$ is immediate by the rule on given statements. If $a \not\prec b$ is in Σ then we need to show that $\Sigma \not\vdash a \prec b$. (This requires a little care: it is not the same thing as saying $\Sigma \vdash a \not\prec b$.) But suppose not, i.e. suppose $\Sigma \vdash a \prec b$. Then we already have $a \not\prec b \in \Sigma$ so from $\Sigma \vdash a \not\prec b$ and the contradiction rule we have $\Sigma \vdash \bot$, which is impossible by the consistency of Σ. $\square$

We already introduced a semantic notion of deduction, $\Sigma \models \sigma$. This semantic deduction is deduction from 'total mathematical knowledge'. If we knew everything there is to know about all possible orderings $<$ on X we could decide questions such as whether $\Sigma \models \sigma$.

The Completeness Theorem, in its alternative form, says that anything that might have been deduced from 'total mathematical knowledge' by examining all of the posets on X can in fact be deduced by a finite formal derivation. In other words our rules of derivation are complete in the sense that they capture all possible (semantic) deductions.

Theorem 4.11 (Contrapositive of the Completeness Theorem) *Suppose X is a set, Σ is a set of statements about a partial order on X, and σ is a further* statement such that $\Sigma \vDash \sigma$. *Then there is a formal proof or derivation from the rules given that Σ implies σ, i.e. $\Sigma \vdash \sigma$.*

Proof If not, $\Sigma \nvdash \sigma$, so by the Reductio Ad Absurdum Rule, $\Sigma \cup \{\sigma'\}$ is consistent, where σ' is the opposite, or negation, of σ. This is because if $\Sigma \cup \{\sigma'\} \vdash \perp$ then $\Sigma \vdash \sigma$ by Reductio Ad Absurdum. It follows that by the first form of the Completeness Theorem there would be a partial order $<$ on X making Σ true and σ false, contradicting the assumption $\Sigma \vDash \sigma$. □

The reader should note in the proof of the theorem just given the example of an informal nested proof by contradiction, i.e. one argument by contradiction inside another.

Notice also that the Soundness Theorem as originally stated is the exact converse of this. It says that $\Sigma \vdash \sigma$ implies $\Sigma \vDash \sigma$.

We have now proved our Completeness and Soundness Theorems for this system. But so far in this chapter, despite the new notation and definitions and care required in separating the objects of discussion (formal derivations) from our proofs about them, the mathematics has been comparatively straightforward and no particularly interesting mathematical results have been proved. This changes with the next theorem, which relies on the following definition and lemma.

Definition 4.12 Say a statement is *positive* if it is of the form $a \prec b$ and *negative* if it is of the form $a \nprec b$.

Lemma 4.13 *Suppose X is a non-empty set, and Σ is a consistent set of positive statements about a partial order on the elements of X. Suppose a, b are in X and are distinct. Then either $\Sigma \cup \{a \prec b\}$ is consistent or $\Sigma \cup \{b \prec a\}$ is consistent.*

Proof Let us suppose that $\Sigma \cup \{b \prec a\}$ is not consistent, so by the Reductio Ad Absurdum Rule $\Sigma \vdash b \nprec a$. Then by the Completeness Theorem and the consistency of Σ there is a partial order $<$ on X making Σ true, and by the Soundness Theorem the order $<$ must also make $b \nprec a$.

We now define a new order $\ll$ on X as follows:

$$x \ll y \text{ if and only if either } x < y \text{ or both } x \leqslant a \text{ and } b \leqslant y.$$

We need to show that this is a poset that makes all of Σ true and also makes $a \prec b$ true.

To check transitivity, suppose that $x \ll y \ll z$. There are several cases.

Firstly, if $x < y < z$ we have $x < z$ by transitivity of $<$ so $x \ll z$. Secondly, if $x < y$ and $y \leqslant a$ and $b \leqslant z$ then $x \leqslant a$ and $b \leqslant z$ by transitivity so $x \ll z$. Thirdly, if $x \leqslant a$ and $b \leqslant y$ and $y \leqslant a$ and $b \leqslant z$ then $b \leqslant a$ which is false by assumption, so this case does not occur. Finally, if $x \leqslant a$ and $b \leqslant y$ and $y < z$ then $x \leqslant a$ and $b \leqslant z$ so $x \ll z$. Therefore $\ll$ has the transitivity property.

Now suppose $x \ll x$. This would imply one of: $x < x$, contradicting irreflexivity of $<$; or $x \leqslant a$ and $b \leqslant x$, implying $b \leqslant a$, another contradiction. Therefore $\ll$ has the irreflexivity property.

Note too that $a \leqslant a$ and $b \leqslant b$ so $a \ll b$.

Finally note that every statement in Σ is of the form $c \prec d$ and true in the poset X with $<$. So by construction of $\ll$ we also have $c \ll d$. In other words, $\ll$ also makes Σ true.

So this shows that there is a partial order $\ll$ on X that makes Σ true and also makes $a \prec b$ true. Therefore $\Sigma \nvDash a \nprec b$ and so by the contrapositive of the Soundness Theorem $\Sigma \nvdash a \nprec b$ hence $\Sigma \cup \{a \prec b\}$ is consistent by Reductio Ad Absurdum. $\qquad\square$

Theorem 4.14 (Any partial order may be linearised) *Let $<$ be a partial order on a non-empty set X. Then there is a linear order $\ll$ on X such that $x < y$ implies $x \ll y$.*

Proof Let Σ_0 be the set of all positive statements about the partial order $<$ on X. Let Y be the set of all consistent sets $\Sigma \supseteq \Sigma_0$ of positive statements of the form $x \prec y$. We order Y using the usual set-inclusion order, $\subseteq$. Then Y has the Zorn property since for any chain $C \subseteq Y$ of such Σ the union of this chain $\bigcup C = \{\sigma : \sigma \in \text{ some } \Sigma \in Y\}$ contains Σ_0 and is consistent. For the latter, note that if $\bigcup C \vdash \bot$ then this says there is a proof of $\bot$ from $\bigcup C$. This proof is a finite object hence uses only finitely many elements from $\bigcup C$. All of these will be in some $\Sigma \in C$ since C is a chain.

Therefore, by Zorn's Lemma, there is a maximal element Σ^+ of Y, and by the Completeness Theorem there is some order $\ll$ on X making Σ^+ true. But for each $a, b \in X$ either $a \prec b$ or $b \prec a$ is in Σ^+, since by the lemma one of $\Sigma^+ \cup \{a \prec b\}$ or $\Sigma^+ \cup \{b \prec a\}$ is consistent. If $\Sigma^+ \cup \{a \prec b\}$ is consistent then it cannot be a proper extension of the maximal set Σ^+, so $a \prec b \in \Sigma^+$. Similarly for the other case. Therefore the order $\ll$ making Σ^+ true is a linear order. It also makes all statements in Σ_0 true, since $\Sigma_0 \subseteq \Sigma^+$ and it makes everything in Σ^+ true. Thus $x < y$ implies $x \prec y \in \Sigma_0$ which implies $x \ll y$, as required. $\qquad\square$

We finish this section with another consequence of Lemma 4.13, which is more 'logical' in nature. A general theme in many parts of logic is as follows: because formal proofs are rather special and difficult to come by, we should feel that having a *proof* of a statement is often mathematically rather stronger than just knowing that the statement is true. This is similar to the fact that knowing a particular infinite path in a tree is often more useful than simply knowing that the tree is infinite. This observation is particularly relevant when the statement is provable from some 'weakened' form of the system, since one then has to work even harder to get the formal proof. If this intuition is correct, there should be some nice mathematical consequences when we know a formal proof exists, rather than just knowing the statement is true, and sometimes the proof itself can be converted to a proof of something better.

Our last result in this section is a rather pretty illustration of exactly this situation. It says that if a negative statement can be proved from a set of positive statements then there is a stronger positive statement we could have proved instead. Of course this only works for the particular system of this chapter, and the reason it works is Lemma 4.13. The proof uses both Soundness and Completeness Theorems.

Theorem 4.15 *If Σ is a set of positive statements about a partial order $<$ on a set X and $a, b \in X$ are distinct, then $\Sigma \vdash b \not\prec a$ implies that $\Sigma \vdash a \prec b$.*

Proof We show the contrapositive, that $\Sigma \not\vdash a \prec b$ implies $\Sigma \not\vdash b \not\prec a$.

Suppose $\Sigma \not\vdash a \prec b$. Then $\Sigma \cup \{a \not\prec b\} \not\vdash \bot$ by the Reductio Ad Absurdum Rule. So by the Completeness Theorem there is a partial order on X making Σ and $a \not\prec b$ true. By the proof of Lemma 4.13, this shows that there is a partial order $\ll$ on X making Σ true and also making $b \prec a$ true. Hence by the Soundness Theorem $\Sigma \cup \{b \prec a\} \not\vdash \bot$ and $\Sigma \not\vdash b \not\prec a$ by the contradiction rule, as if $\Sigma \vdash b \not\prec a$ then $\Sigma \cup \{b \prec a\} \vdash \bot$. $\qquad\square$

Of course, no such result can hold without some special condition on Σ such as it being a set of positive statements. Indeed $\{b \not\prec a\} \vdash b \not\prec a$ always holds, but $\{b \not\prec a\} \not\vdash a \prec b$.

4.2 Examples and exercises

We have introduced a lot of notation and terminology in this chapter, and some of it is quite tricky to master at first. For example, there is a big difference between $\Sigma \not\vdash a \prec b$ and $\Sigma \vdash a \not\prec b$. Some of the arguments employed are a little subtle too, with uses of contrapositive statements and also uses of informal proofs by contradiction within a larger argument. Some of the exercises here investigate these areas further.

The main point of the chapter is to illustrate the key methods of logic, of formal systems and semantics, applied to a familiar situation. Once again, the fact that proofs are finite objects is used in several places, for example to show that a poset of a consistent set of sentences has the Zorn property (in the proof of the theorem that any partial order may be linearised). The technique of switching between the point-of-view of formal proofs, or $\vdash$, and what these formal proofs mean, or $\vDash$, using the Completeness and Soundness Theorems gives useful and interesting mathematical information.

Exercise 4.16 Show that $\varnothing \vdash \{a \not\prec a\}$.

Exercise 4.17 Show that:

 (i) $\{a \prec b, b \not\prec c\} \not\vdash a \not\prec c$
 (ii) $\{a \not\prec b, b \prec c\} \not\vdash a \not\prec c$
 (iii) $\{a \not\prec b, b \not\prec c\} \not\vdash a \not\prec c$

(Hint: use soundness.)

Exercise 4.18 Explain the difference between the two statements $\Sigma \not\vdash a \prec b$ and $\Sigma \vdash a \not\prec b$. Does either one of these imply the other? Justify your answer.

Exercise 4.19 Suppose Σ is a consistent set of statements about elements of a set X, and $a, b \in X$. Prove that either $\Sigma \cup \{a \prec b\}$ is consistent or $\Sigma \cup \{a \not\prec b\}$ is consistent.

Exercise 4.20 Suppose Σ is a set of negative statements (i.e. of statements of the form $a \not\prec b$ for $a, b \in X$) and suppose σ is positive. Show that $\Sigma \not\vdash \sigma$.

The proof we gave of the Completeness Theorem (Theorem 4.10) was rather specialised to partial orders and does not generalise nicely to other systems. The following exercise gives an alternative.

Exercise 4.21 Suppose Σ is a consistent set of statements about elements of a set X. Show that there is $\Sigma^+ \supseteq \Sigma$ which is maximal in $X = \{\Gamma \supseteq \Sigma : \Gamma \not\vdash \bot\}$, where Γ ranges over all sets of statements of the form $a \prec b$ or $a \not\prec b$ about elements of X. Define $<$ on X by $a < b$ if and only if $\Sigma^+ \vdash a \prec b$. Show that $<$ is a partial order on X such that Σ^+ is precisely the set of statements true for $<$.

To illustrate this, the following exercise discusses a modification of our system to allow for deductions about linear orders.

Exercise 4.22 Add a new rule to the system that says, if a, b are distinct elements of X such that the statement $a \not\preccurlyeq b$ has been deduced, then $b \prec a$ may also be deduced. Also, redefine $\Sigma \vDash \sigma$ to mean every *linear order* making Σ true must make σ true. State and prove Completeness and Soundness Theorems for this system. (Hint: for the proof of a Completeness Theorem, follow the idea of the previous exercise.)

Similar systems can describe preorders and equivalence relations, as the next two exercises show.

Exercise 4.23 Let X be a set with at least two elements. Consider the following formal system, where statements are allowed to be of the form $a \preccurlyeq b$ where $a, b \in X$.

The rules of deduction are: (1) the 'given statements rule' above; (2) for an $a \in X$ the statement $a \preccurlyeq a$ can be deduced; and (3) if $a \preccurlyeq b$ and $b \preccurlyeq c$ have been deduced where $a, b, c \in X$ then we may deduce $a \preccurlyeq c$.

Prove the following Soundness Theorem that if $\Sigma \vdash a \preccurlyeq b$ then $a \leqslant b$ holds in any preorder on X that makes Σ true. Prove also that '$\Sigma \vdash a \preccurlyeq b$' defines a preorder on X that makes Σ true.

Exercise 4.24 Develop a formal system for proofs about an equivalence relation $\sim$ on a set X. Prove Soundness and Completeness Theorems for your system.

4.3 Linearly ordering algebraic structures*

In general, the notion of formal proof and the Completeness Theorem can be regarded as a clever way of organising a poset with the Zorn property and the consequences of having maximal elements. So in principle, the logical techniques described in this book can be avoided. That said, in many real-life situations the correct notion of consistency required is so complicated that it seems humanly impossible to avoid some of the logical manoeuvres described in later chapters.

Exercise 4.25 Prove the theorem that any partial order can be linearised by using Zorn's Lemma directly.

As an astute reader will have observed, some of the results in this chapter can be proved without the use of the Axiom of Choice, including (unusually) the Completeness Theorem. On the other hand, the theorem that any partial order can be linearised requires some Choice, but it turns out that it is not equivalent to the full Axiom of Choice. In the rest of this section we shall look at other examples where these systems of proof can help organise and prove some results in algebra related to ordered structures; all of these will require Zorn's Lemma or the Axiom of Choice in some way.

We start with abelian groups. Recall that an *abelian group* is a group G with binary operation $+$ and identity 0 for which the operation is commutative: $x + y = y + x$ for all $x, y \in G$. We write nx for $x + x + \cdots + x$ where x appears n times.

Definition 4.26 An *ordered abelian group* is an abelian group G with a linear order $<$ on G such that

$$x < y \text{ implies } x + z < y + z$$

for all $x, y, z \in G$.

An abelian group G is *orderable* if there is some linear order $<$ on G making G into an ordered abelian group.

For example, the group $(\mathbb{Z}, +)$ is orderable, and in fact the usual order satisfies the axiom above. On the other hand, no finite cyclic group is orderable since if x is a generator and $0 < x$ then the axiom above implies $0 < x < x + x < x + x + x < \cdots$ but $nx = x + \cdots + x = 0$ where n is the order of x. This contradicts transitivity of $<$. A similar argument applies when $x < 0$.

Definition 4.27 An element x of an abelian group G is *torsion* if $x \neq 0$ and there is $n > 0$ in $\mathbb{N}$ such that $nx = 0$. A group G is *torsion-free* if there are no torsion elements of G other than the identity element, 0.

Exercise 4.28 Continue the argument for finite cyclic groups above, showing that an abelian group with a torsion element x is not orderable.

We want to investigate which abelian groups are orderable. The first step is to extend the system of Exercise 4.22 and prove Completeness and Soundness Theorems.

Let our set X of elements be the elements of G, an abelian group. Add a new rule to the system for linear orders in Exercise 4.22 that says that if $a, b, c \in X$ and $a \prec b$ has been deduced then $a' \prec b'$ may be deduced, where $a' = a + c$ and $b' = b + c$.

Exercise 4.29 Prove that $\Sigma \nvdash \perp$ if and only if G is orderable with a linear order $<$ making all statements in Σ true; in particular $\varnothing \nvdash \perp$ if and only if G is orderable. (For completeness, use a Zorn's Lemma argument similar to that in Exercise 4.21.)

The next stage is to simplify this proof system so that we can analyse it algebraically. Using the group operation we can rewrite the statement $a \prec b$ as $0 \prec c$ where $c = b - a \in G$. Because our orders will be linear, $a \not\prec b$ is 'the same as' $b \prec a$ for $a \neq b$, and we will have no need for negative statements. So the new simplified system will have statements $\perp$ and $0 \prec a$ for $a \in G$ only. The new proof rules will be:

- the Given Statements Rule;
- the rule, from $0 \prec 0$ deduce $\perp$;
- the rule, from $0 \prec a$ and $0 \prec b$ deduce $0 \prec c$, where $c = a + b$;
- the Reductio Ad Absurdum Rule in the form, if from assumption $0 \prec a$ (where $a \neq 0$) you can prove $\perp$, then you may deduce $0 \prec b$ without any assumptions, where $b = -a$ is the inverse of a.

Note that the new Reductio Ad Absurdum Rule combines the old Reductio Ad Absurdum Rule with the rule for linearity of the order. The rule about $a + b$ is a special case of the rule about the order respecting addition in the group.

We could prove new Completeness and Soundness Theorems for the new simplified system directly. However, in this case there is a simpler alternative: to define a translation of one system into the other.

Definition 4.30 We define a translation t of the system of Exercise 4.29 into the new system by letting the translation $t(\perp)$ of $\perp$ be $\perp$, $t(a \prec b)$ be $0 \prec b - a$, and for $a \neq b$, we let $t(a \not\prec b)$ be $0 \prec a - b$. (For $a = b$ the statement $a \not\prec b$ is always provable and may be safely ignored.)

Exercise 4.31 Prove by induction on proofs that $a \prec b$ is provable in the first system from Σ implies that $t(a \prec b)$ is provable in the simplified system from $\{t(\sigma) : \sigma \in \Sigma\}$. Also, arguing directly this time, prove a converse to this statement.

Hence deduce Completeness and Soundness Theorems for the new system, including the statement that $\Sigma \not\vdash \perp$ if and only if the group G is orderable in such a way that every statement in Σ is true for the new order on G.

We are now in a position to prove our converse to Exercise 4.28 and prove that every torsion-free abelian group is orderable. The main technical result (giving the induction on length of proofs that is required) is the following.

Proposition 4.32 *Suppose G is torsion-free and abelian, $x_1, \ldots, x_k \in G$ are all non-zero, and*

$$\{0 \prec x_1, 0 \prec x_2, \ldots, 0 \prec x_k\} \vdash 0 \prec y$$

for some $y \in G$. Then there are $n_i \in \mathbb{N}$, not all zero, and $m \in \mathbb{N}$ such that

$$my = \sum_{i=1}^{k} n_i x_i.$$

Proof By induction on the length of proofs. We consider a proof of

$$\{0 \prec x_1, 0 \prec x_2, \ldots, 0 \prec x_k\} \vdash 0 \prec y$$

and assume that the statement is true for all $x_1, \ldots, x_l, y \in G$ and all shorter proofs.

The base case is when the 'given statements' rule is used to deduce $0 \prec y$ and so $y = x_j$ for some j. Then $y = \sum_{i=1}^{k} n_i x_i$ where $n_j = 1$ and all other $n_i = 0$.

For a proof in which the last step is not obtained using the 'given statements' rule, by examining the rules, we see that our proof can take one of only two forms. Either the last step is the rule 'from $0 \prec u$ and $0 \prec v$ deduce $0 \prec u + v$', or the last step is an instance of Reductio Ad Absurdum. Considering the first of these we have $y = u + v$ where $u, v \in G$ and shorter proofs of $\{0 \prec x_1, 0 \prec x_2, \ldots, 0 \prec x_k\} \vdash 0 \prec u$ and $\{0 \prec x_1, 0 \prec x_2, \ldots, 0 \prec x_k\} \vdash 0 \prec v$ and so by induction there are $n_i' \in \mathbb{N}$, not all zero, $n_i'' \in \mathbb{N}$, not all zero, and $p, q \in \mathbb{N}$ such that $pu = \sum_{i=1}^{k} n_i' x_i$, and $qv = \sum_{i=1}^{k} n_i'' x_i$. Now let m be the lowest common multiple of p, q, $rp = sq = m$. If $p = 0$ let $m = q$ and $r = s = 1$, and if $q = 0$ let $m = p$ and $r = s = 1$. In each case, note that $my = rpu + sqv = \sum_{i=1}^{k}(rn_i' + sn_i'')x_i$ and as $r, s > 0$ if either n_i' or n_i'' is non-zero, then so is $rn_i' + sn_i''$, as required.

Now suppose the last step of the proof is an instance of Reductio Ad Absurdum, so that $y \neq 0$ and the proof looks like

Formal proof

$\ldots$	(1)	
$0 \prec -y$	(2)	Assumption
$\ldots$	(3)	
$0 \prec 0$	(4)	
$\perp$	(5)	Contradiction
$0 \prec y$	(6)	Reductio Ad Absurdum

Then the conclusion $0 \prec 0$ in line 4 follows from the given statements $0 \prec x_1$,

$0 \prec x_2, \ldots, 0 \prec x_k$ and the additional assumption $0 \prec -y$, and thus by our induction hypothesis there are $m, n_i \in \mathbb{N}$ (not all zero) such that

$$0 = m(-y) + \sum_{i=1}^{k} n_i x_i.$$

Now $y \neq 0$ (otherwise the assumption to the Reductio Ad Absurdum Rule is not allowed) so if all of the n_i are zero we would have $my = 0$ contradicting the fact that G is torsion-free. Thus at least one of the n_i is non-zero and

$$my = \sum_{i=1}^{k} n_i x_i$$

as required. ☐

Corollary 4.33 *Let G be torsion-free and abelian. Then G is orderable.*

Proof If not then $\perp$ is provable from no given statements. Without any statements previously given the first step must be a subproof with some assumption $0 \prec y$ ($y \neq 0$), and the conclusion to this subproof is the contradiction $0 \prec 0$. Thus by the proposition $0 = ny$ for some $n \in \mathbb{N}$, n not zero. Thus G is not torsion-free. ☐

We now turn, briefly, to non-abelian groups, i.e. groups where the commutativity axioms might fail. As is conventional, we revert to multiplicative notation. Since the group may no longer be abelian we must specify the side that the action in the axiom '$a < b$ implies $ac < bc$' takes place.

Definition 4.34 A *right-ordered group* is a group G with a linear order $<$ on G such that $a < b$ implies $ac < bc$ for all $a, b, c \in G$. A *right-orderable group* is a group G which may be made into a right-ordered group by the choice of an appropriate order.

It is not so easy to give a purely algebraic criterion that describes when a group G can be ordered in such a way that it becomes a right-ordered group. However, the analogue of Exercise 4.29 holds.

Exercise 4.35 Let X be the set of elements of a group G. Add a new rule to the system for linear orders in Exercise 4.22 that says that if $a, b, c \in X$ and $a \prec b$ has been deduced then $a' \prec b'$ may be deduced, where $a' = ac$ and $b' = bc$. Prove that $\Sigma \nvdash \perp$ if and only if there is a linear order on X making G into a right-ordered group in which all statements in Σ are true, and in particular $\varnothing \nvdash \perp$ if and only if G is right-orderable.

The following exercise gives a family of examples that are right-orderable. In attempting it, it may be helpful to know that any set such as Ω has a *well-order*, i.e. a linear order in which every non-empty subset has a least element. (This fact about well-orders can be proved from Zorn's Lemma and is in fact equivalent to the Axiom of Choice.)

Exercise 4.36 Let G be a group of permutations of a linearly ordered set Ω, i.e. each $g \in G$ is a bijection $g\colon \Omega \to \Omega$, with the group operation being composition of maps. The order on Ω will be denoted $\ll$. Suppose each $g \in G$ respects $\ll$ in the sense that $x \ll y$ implies $g(x) \ll g(y)$ for each $x, y \in \Omega$. Show that G is right-orderable.

The 'simplified system' for orderable abelian groups that we looked at above concerns specifications of which elements are positive, i.e. for which a we have $0 < a$. A related result on orderability of fields goes via specifying the positive elements in a field.

Definition 4.37 A field F is an *ordered field* if it has a linear order $<$ such that $x < y$ implies $x + z = y + z$ for all $x, y, z \in F$ and $x < y$ implies $xz = yz$ for all *positive* $x, y, z \in F$. The field F is *orderable* or *formally real* if there is such an order that can be defined on F.

Definition 4.38 Let F be a field. A *pre-positive cone* of F is a set $P \subseteq F$ which is closed under addition and multiplication, contains 1 but does not contain 0, and contains all x^2 for $x \neq 0$ in F. A *positive cone* of F is a maximal pre-positive cone $P \subseteq F$.

Exercise 4.39 Use Zorn's Lemma to show that a pre-positive cone of a field F is always contained in a positive cone.

Exercise 4.40 Let P be a positive cone of F. Define $<$ by $x < y$ if and only if $y - x \in P$. Show that this defines a linear order on F that makes F into an ordered field.

Exercise 4.41 Deduce that F is orderable if and only if F has a pre-positive cone, and that this is the case if and only if no finite sum of non-zero squares from F equals -1.

Exercise 4.42 Reinterpret these results as a formal system, proving Soundness and Completeness Theorems for your system.

5

Boolean algebras

5.1 Boolean algebras

Partially ordered and linearly ordered sets may be interesting but they are not really 'logic' in the usual sense of the word: they do not represent logical statements nor do they model logical operations such as 'not', 'and', or 'or'. We arc now going to investigate special kinds of posets called boolean algebras, the elements of which can be used to represent logical propositions.

We start by adding extra operations such as 'and', 'or', and 'not' to a poset, turning it into a boolean algebra. (But it turns out that it is much more convenient to look at non strict posets with an order relation $\leqslant$, though you may take a strict partial order and change it to a non strict order as explained before.) The axioms for a boolean algebra are given in three groups, making up the axioms for a lattice, a distributive lattice and a boolean algebra proper. We give each group separately.

Definition 5.1 A *lattice* is a poset X with non-strict order $\leqslant$ such that every pair of elements x, y of X has a *greatest lower bound* denoted $\inf(x, y)$ or $x \wedge y$ satisfying

$$x \wedge y \leqslant x \text{ and } x \wedge y \leqslant y$$

and also

$$z \leqslant x \text{ and } z \leqslant y \text{ imply } z \leqslant x \wedge y$$

for all $z \in X$; furthermore, every pair of elements x, y of X has a *least upper bound* denoted $\sup(x, y)$ or $x \vee y$ satisfying

$$x \vee y \geqslant x \text{ and } x \vee y \geqslant y$$

and also

$$z \geqslant x \text{ and } z \geqslant y \text{ imply } z \geqslant x \vee y$$

55

for all $z \in X$.

Exercise 5.2 Let X be a poset. Suppose $a, b \in X$ are both greatest lower bounds of $x, y \in X$, i.e. suppose that $a \leqslant x$, $a \leqslant y$, $b \leqslant x$ and $b \leqslant y$ (so both a, b are lower bounds of x, y) and that whenever z satisfies $z \leqslant x$ and $z \leqslant y$ then $z \leqslant a$ and $z \leqslant b$ (so both a, b are greatest lower bounds of x, y). Show that $a = b$. Thus there can be at most one greatest lower bound $x \wedge y$ of $x, y \in X$. Similarly, show that there can be at most one least upper bound $x \vee y$ of $x, y \in X$.

Exercise 5.3 Show that any linearly ordered set is a lattice.

Exercise 5.4 Show that any finite poset is a lattice.

Exercise 5.5 Find an infinite poset which is not a lattice.

Definition 5.6 A *distributive lattice* is a lattice X such that the distributivity axioms hold for $\wedge$ and $\vee$, i.e.

$$x \wedge (y \vee z) = (x \wedge y) \vee (x \wedge z)$$

and

$$x \vee (y \wedge z) = (x \vee y) \wedge (x \vee z)$$

for all $x, y \in X$.

Example 5.7 If X is an infinite set, the set F of all finite subsets of X forms a distributive lattice under $\subseteq$.

Example 5.8 Let $A = \{0, 1, 2, 3, 4\}$ be the poset where $0 < 1 < 2 < 4$, $0 < 3 < 4$ and no other relations between these elements hold. Then A is a non-distributive lattice. (Exercise: compute $1 \wedge (2 \vee 3)$ and $(1 \wedge 2) \vee (1 \wedge 3)$ in A.)

Example 5.9 Let $B = \{0, 1, 2, 3, 4\}$ be the poset where $0 < 1 < 4$, $0 < 2 < 4$, $0 < 3 < 4$ and no other relations hold. Then B is also a non-distributive lattice. (Exercise: compute $1 \wedge (2 \vee 3)$ and $(1 \wedge 2) \vee (1 \wedge 3)$ in B.)

Remark 5.10 The last two examples are typical of non-distributive lattices: in fact, every non-distributive lattice contains a copy of either A of Example 5.8 or B of Example 5.9. An ambitious reader might try to prove this as a more difficult exercise.

Definition 5.11 A *boolean algebra* is a distributive lattice X containing two special elements denoted $\perp$, $\top$ (also sometimes denoted 0, 1, or F, T, respectively) and having a function $X \to X$ called *complementation* or *negation* written $'$ (or $\neg$ or ^{c}) such that:

(i) $\perp$ is the minimum element of X and $\top$ is the maximum element of X, i.e. $\perp \leqslant x$ and $x \leqslant \top$ for all $x \in X$;

(ii) for all $x \in X$ the complement x' of x satisfies $x \wedge x' = \perp$ and $x \vee x' = \top$.

Example 5.12 The poset $2 = \{\top, \perp\}$ with $\perp < \top$ is a boolean algebra when we define $\top' = \perp$, and $\perp' = \top$. You can check that $\perp \wedge \perp = \perp \wedge \top = \top \wedge \perp = \perp \vee \perp = \perp$ and $\top \wedge \top = \perp \vee \top = \top \vee \perp = \top \vee \top = \top$ all hold, as expected.

Example 5.13 If X is a non-empty set, the poset $P(X)$ of all subsets of X with the usual $\subseteq$ is a boolean algebra where $\perp = \varnothing$, $\top = X$, $A' = X \setminus A$ (set complementation) and $A \wedge B = A \cap B$, $A \vee B = A \cup B$.

Example 5.14 The poset with a single element (simultaneously called both $\top$ and $\perp$) is, according to the axioms given above, a boolean algebra, called the improper or *degenerate algebra*. This algebra is not particularly interesting or useful, but does come up occasionally, e.g. when we are studying contradictory systems.

The following proposition gives a useful alternative list of properties of $\wedge$, $\vee$ and $\top$, $\perp$ in a boolean algebra.

Proposition 5.15 *The following hold for all elements a, b, c in a boolean algebra X*

(i) $a \wedge a = a \vee a = a$

(ii) $a \wedge b = b \wedge a$ *and* $a \vee b = b \vee a$

(iii) $a \wedge (b \wedge c) = (a \wedge b) \wedge c$ *and* $a \vee (b \vee c) = (a \vee b) \vee c$

(iv) $a \wedge (a \vee b) = a \vee (a \wedge b) = a$

(v) $a \wedge (b \vee c) = (a \wedge b) \vee (a \wedge c)$ *and* $a \vee (b \wedge c) = (a \vee b) \wedge (a \vee c)$

(vi) $\perp \wedge a = \perp$, $\perp \vee a = a$, $\top \wedge a = a$ *and* $\top \vee a = \top$

(vii) $a \wedge a' = \perp$ *and* $a \vee a' = \top$.

Additionally, we have alternative characterisations of the order relation in a boolean algebra. For all a, b the following are equivalent: $a \leqslant b$; $a \wedge b = a$; $a \vee b = b$; $b' \wedge a = \perp$; and $a' \vee b = \top$.

Proof The first three parts are easy and are left as an exercise.

For part (iv), note that $a \vee b \geqslant a$ so a is the greatest lower bound of $a \vee b$, a, i.e. $a \wedge (a \vee b) = a$. Similarly a, $a \wedge b \leqslant a$ so $a \vee (a \wedge b) = a$.

The distributivity laws were already given as axioms of boolean algebras so there is nothing to prove in the next point.

Since $\perp$ is the least element, it must be the greatest lower bound of itself and any other $a \in X$. Hence $\perp \wedge a = \perp$. The other three equations in this group are similar.

The statements $a \wedge a' = \perp$ and $a \vee a' = \top$ were also given as axioms for boolean algebras.

If $a \leqslant b$ then a is a lower bound for b and hence the greatest lower bound for a, b together. Conversely if $a \wedge b = a$ then a is a lower bound for b so $a \leqslant b$. The next case is similar. If $a' \vee b = \top$ then doing '$a \wedge$' to both sides and using distributivity we have the left hand side is equal to

$$a \wedge (a' \vee b) = (a \wedge a') \vee (a \wedge b) = \perp \vee (a \wedge b) = a \wedge b$$

and this equals the right hand side, which is $a \wedge \top = a$, so $a \wedge b = a$. Conversely doing '$a' \vee$' to both sides of $a \wedge b = a$ and using distributivity gives $a' \vee b = \top$. The other case is similar. □

The last part of this proposition shows that the order relation in a boolean algebra can be defined in terms of $\wedge$ or $\vee$. This gives us another way of specifying a boolean algebra as an algebraic structure somewhat like a ring, with operations $\wedge, \vee, '$ and elements $\top, \perp$ and defining $\leqslant$ from these.

Proposition 5.16 *Let* $(X, \wedge, \vee, ', \top, \perp)$ *be an algebraic structure satisfying properties* (i)–(vii) *in Proposition 5.15 for all* $a, b, c \in X$. *Define* $\leqslant$ *on* X *by any of the clauses given in the last part of the proposition. Then this makes* X *into a boolean algebra, with the sup and inf operations being* $\vee$ *and* $\wedge$, *and with maximum and minimum elements* $\top$ *and* $\perp$.

Proof It is necessary to check that $\leqslant$ is a partial order satisfying the axioms in Definition 5.11. Left as an exercise. □

We can thus swap between our view of a boolean algebra as a poset or as an algebraic object like a ring at will, just like we swap between our views of a strict and non-strict poset.

Purists would read $\top, \perp$ as 'top' and 'bottom', respectively, and $\wedge, \vee$ as 'meet' and 'join'. However, the symbols $\top, \perp$ are often read as 'true' and 'false', and $\wedge, \vee$ as 'and' and 'or'. 'Or' is always inclusive or, i.e. meaning one or the other *or both*. The complementation operation is 'not', especially when it is written as $\neg$.

The mathematically rather common implication, 'if a then b', often causes some consternation, since from the point of view of natural language we normally expect this to be true only when a and b have something to do with each other. (So 'if the sky has dark clouds then it will rain' seems quite reasonable, but 'if there are small green men on Mars then it is raining' seems as if it should be false, because what have the strange creatures on Mars got to do with our rain? But on the other hand, if it really is raining, surely the statement is true whatever might be on Mars? And, if there are no small green men on Mars, is the statement not vacuously true anyway?) There is no mathematical symbol that captures exactly this 'natural' idea of 'if a then b' where a should be relevant to b, but the $\leqslant$ in the boolean algebra does quite a good job in all other respects. Think of $a \leqslant b$ as saying that b is at least as true as a. Then $a \leqslant b$ is quite a good *mathematical* definition for 'if a then b', even if it does not capture all the nuances of natural language.

By the last part of Proposition 5.15, $a \leqslant b$ is equivalent in a boolean algebra to the statement that $a' \vee b = \top$, or in words 'not-a or b'.

We investigate this order relation as 'implies' or 'if... then...' in the next few propositions. The first of these says that if a implies b then a and c implies b and c, and with something similar for 'or'.

Proposition 5.17 *Let B be a boolean algebra with $a, b, c \in B$ and $a \leqslant b$. Then $a \wedge c \leqslant b \wedge c$ and $a \vee c \leqslant b \vee c$.*

Proof For $a \wedge c \leqslant b \wedge c$ it suffices to observe that $a \wedge c \leqslant a \leqslant b$ and $a \wedge c \leqslant c$ so $a \wedge c$ is a lower bound of both b, c hence is less than or equal to the least such lower bound.

For $a \vee c \leqslant b \vee c$ argue similarly using $a \leqslant b \leqslant b \vee c$ and $c \leqslant b \vee c$. $\square$

The next proposition is the law of contraposition: the *contrapositive* of 'a implies b' is 'not-b implies not-a'. As shown here, an implication and its contrapositive are equivalent.

Proposition 5.18 *Let B be a boolean algebra and suppose $a \leqslant b$. Then $b' \leqslant a'$.*

Proof If $a \leqslant b$ then

$$b' = b' \wedge (a' \vee a) \leqslant b' \wedge (a' \vee b) = (b' \wedge a') \vee (b \wedge b') = b' \wedge a'$$

so $b' \leqslant a'$. $\square$

Proposition 5.19 *Let B be a boolean algebra and suppose $b' \leqslant a'$. Then $a \leqslant b$.*

Proof Exercise, to be done in a similar way. Or use the previous result and Proposition 5.22 on Uniqueness of Complements below, where it is shown that $x'' = x$ in any boolean algebra B. □

The next proposition is often useful. It says that to determine whether a statement a holds it suffices to consider two cases: one where b is true, and one where it is false. Of course b can be any proposition at all. (It can even be a itself, in which case the proposition would be a triviality.)

Proposition 5.20 *Let B be a boolean algebra and suppose $a, b \in B$. Then $(a \wedge b) \vee (a \wedge b') = a$.*

Proof By the distributivity laws,

$$(a \wedge b) \vee (a \wedge b') = a \wedge (b \vee b') = a \wedge \top = a$$

as required. □

There is a similar dual statement, with 'and' and 'or' swapped over.

Proposition 5.21 *Let B be a boolean algebra and suppose $a, b \in B$. Then $(a \vee b) \wedge (a \vee b') = a$.*

Proof Exercise. □

Finally, we have an important result concerning 'not'.

Proposition 5.22 (Uniqueness of Complements) *Let B be a boolean algebra, with $a, b \in B$, and suppose that $a \wedge b = \bot$ and $a \vee b = \top$. Then $b = a'$. In particular $x'' = x$ for all $x \in B$.*

Proof Assume that $a \wedge b = \bot$ and $a \vee b = \top$. Then

$$a' = a' \vee \bot = a' \vee (a \wedge b) = (a' \vee a) \wedge (a' \vee b) = \top \wedge (a' \vee b) = a' \vee b$$

by distributivity, so $a' \geqslant b$, and similarly

$$a' = a' \wedge \top = a' \wedge (a \vee b) = (a' \wedge a) \vee (a' \wedge b) = \bot \vee (a' \wedge b) = a' \wedge b$$

so $a' \leqslant b$. Hence $a' = b$.

Now let a in the previous paragraph be x', and put $b = x$. Then $x' \wedge b = \bot$ and $x' \vee b = \top$, hence $b = x''$. □

5.2 Examples and exercises

Exercise 5.23 Carry out the procedure that starts with a boolean algebra as an algebraic structure $(X, \wedge, \vee, ', \top, \bot)$ satisfying the statements in Proposition 5.15, defining an order relation by $a \leqslant b$ if and only if $a \wedge b = a$ and then verifying that the poset obtained has the properties in Definition 5.11.

Exercise 5.24 Show that in any boolean algebra the expressions $(a \wedge b)'$ and $(a' \vee b')$ are equal.

Exercise 5.25 Let B be a boolean algebra. Show that the complementation operation $': B \to B$ is one-to-one and onto.

Exercise 5.26 The following statements are all intuitively correct. (To verify that they are valid, just multiply the first equation by 2 or 0.)

(a) $1 = 1$ implies $2 = 2$; (b) $1 = 2$ implies $2 = 4$; (c) $1 = 2$ implies $0 = 0$.

On the other hand the following is not a correct implication:

(d) $1 = 1$ implies $2 = 4$.

Assuming that a mathematical implication a implies b only depends on the truth of a and b, and writing $\to$ for 'implies', deduce that mathematical implication should be fully determined by the following rules: $(\top \to \top) = \top$; $(\top \to \bot) = \bot$; $(\bot \to \top) = \top$; $(\bot \to \top) = \top$.

5.3 Boolean algebra and the algebra of Boole*

The name 'boolean algebra' celebrates the contributions of George Boole to logic. In fact, Boole's work is so important that the word 'boolean' has entered the language and is conventionally spelled without a capital letter. Boole was the first to employ algebraic techniques in logic and is rightly commemorated for his contributions to logic, particularly in his books *The Mathematical Analysis of Logic* (1847) and *An Investigation of the Laws of Thought* (1854). However, in keeping with the traditions of the time, Boole never wrote down a full set of axioms for his algebra, and it is not quite clear how one might describe the algebra of Boole in modern terms.

Certainly the symbols $\wedge$ and $\vee$ post-date Boole's work by many years. Boole instead used multiplication and addition operations, in what seems like a ring-like structure. It is commonly believed (especially by those who have not read Boole's books) that Boole worked in what are now called boolean rings.

Definition 5.27 A *boolean ring* is a ring R which is commutative with 1 and for which $x^2 = x$ for all $x \in R$.

Exercise 5.28 Given a boolean ring R, show that by defining $a \wedge b = ab$ and $a \vee b = ab + a(1 - b) + (a - 1)b$ on R, we obtain a boolean algebra.

Exercise 5.29 Given a boolean algebra, we can define a boolean ring on the same set by $ab = a \wedge b$ and $a + b = (a \vee b') \wedge (a' \vee b)$. If this is applied to the boolean algebra resulting from the previous exercise we recover the original addition operation in R.

Exercise 5.30 Show that $x + x = 0$ holds for all x in a boolean ring R. In particular all boolean rings have characteristic 2, i.e. $1 + 1 = 0$.

Exercise 5.31 Let R be a commutative ring with 1 and let $I \subseteq R$ be the ideal generated by all elements $x(1 - x)$ with $x \in R$. Show that R/I is a boolean ring, in fact the largest quotient of R which is a boolean ring.

The operations taking a boolean algebra to a boolean ring and back again described in the last few exercises are inverse to each other. However, Boole's algebra was not that of boolean rings, since (for example) there are a number of places where he carefully includes a coefficient '2' – indeed he sometimes divides by 2 – and yet $1 + 1 = 0$ holds in any boolean ring.

Instead, it seems that Boole's algebra was of a ring R which is commutative and has 1, in which there are no additive or multiplicative nilpotent elements at all (meaning $nx = x + x + \cdots + x \neq 0$ for all $n \neq 0$ and all $x \neq 0$, and $x^n = xx \cdots x \neq 0$ for all n and all $x \neq 0$), but some elements x of R have the (for the time, counter-intuitive) property of being *idempotent elements*, i.e. $x^2 = x$. Boole would call such x *elective elements*.

For an example of such a ring, consider a non-empty set X of individuals and consider R to be the set of functions $f : X \to \mathbb{R}$ with componentwise addition and multiplication, $(f + g)(x) = f(x) + g(x)$ and $(fg)(x) = f(x)g(x)$. Then the *characteristic function* of a subset $A \subseteq R$, defined by $\chi_A(x) = 1$ if $x \in A$ and $\chi_A(x) = 0$ otherwise, is idempotent and represents the set of individuals A. This particular ring can thus be seen to contain a representation of the set $P(X)$ of all subsets of X and the idempotents in it are very closely related to the boolean algebra $P(X)$ with $\subseteq, \cup, \cap$.

The mathematical part of Boole's work in logic can be seen then to be an investigation into the properties of elective, or idempotent, elements of such a ring R, and of functions of such elective arguments. In more modern terms, the set $E \subseteq R$ of electives or idempotents is not a subring of R, but may be made into a boolean ring by redefining the addition operation on it to be $x \oplus y = x + y - 2xy$. The development of boolean algebra from Boole's algebra came some time later.

6

Propositional logic

6.1 A system for proof about propositions

We are going to develop a formal system for proofs about boolean algebras, just as in a previous chapter we developed one for posets. It will also be rich enough to simulate proofs in the systems given in Chapters 3 and 4, though we will not be in a position to explain the precise connections until Chapter 7. (For this, please see Examples 7.27 and 7.29.)

The system will contain objects representing elements of a boolean algebra that say things such as $a \leqslant b$ and $a = b$, but there is a subtle and rather elegant point here: with our extra symbols for $\wedge, \vee, ', \top, \bot$ we do not need to use either of the symbols $<$ or $\leqslant$, since $a \leqslant b$ holds if and only if $a' \vee b = \top$. Instead, the statements in our system will be elements of the boolean algebra – or rather terms *representing* elements in the boolean algebra – and if a statement is provable or derivable we shall think of it as being *true*, or equal to $\top$.

The next definition explains the terms which will represent elements of some boolean algebra.

Definition 6.1 Let X be any set, which for this definition will be called a set of *propositional letters*. The set of *boolean terms* $\mathrm{BT}(X)$ *over X* is defined as a set of expressions, or strings of symbols, from a set of symbols including $(,), \wedge, \vee, \top, \bot, \neg$ and all elements of X, as follows.

- $\bot$ and $\top$ are both boolean terms over X.
- Any element a of X is a boolean term over X.
- If t is a boolean term over X then so is $\neg t$.
- If t and s are boolean terms over X then so are $(t \wedge s)$ and $(t \vee s)$.
- No other object r is a boolean term unless it can be obtained by finitely many applications of the previous four rules.

Example 6.2 If $X = \{a, b, c\}$ then $\top$, a, $((a \wedge \neg b) \vee c)$, $((a \wedge \neg b) \vee c)$, and $\neg \bot$ are all distinct boolean terms over X.

Note that two terms such as $\top$ and $\neg \bot$, or $(a \wedge a)$ and $(a \vee a)$, count as distinct terms since they are different as strings of symbols, even though they will evaluate to the same object in any boolean algebra.

Definition 6.3 Let X be a set, and let $\mathrm{BT}(X)$ be the set of boolean terms over X. A *formal proof* or derivation from assumptions $\Sigma \subseteq \mathrm{BT}(X)$ is a derivation of finite length where each statement in it is an element of $\mathrm{BT}(X)$ and which uses only the following proof rules.

- (Given Statements Rule) Any $\sigma \in \Sigma$ may be deduced from Σ in one step.
- (Top Rule) The statement $\top$ may be deduced from any Σ in one step.
- ($\wedge$-Introduction) If σ and τ have been derived from Σ then $(\sigma \wedge \tau)$ may be deduced from Σ in one further step.
- ($\vee$-Introduction) If τ has been deduced from Σ and $\sigma \in \mathrm{BT}(X)$ then either of $(\sigma \vee \tau)$, $(\tau \vee \sigma)$ may be deduced from Σ in one further step.
- ($\wedge$-Elimination) If $(\sigma \wedge \tau)$ has been deduced from Σ then either σ or τ may be deduced from Σ in one further step.
- ($\vee$-Elimination) If $(\sigma \vee \tau)$ and $\neg \sigma$ have been deduced from Σ then τ may be deduced from Σ in one further step.
- ($\neg$-Elimination) If $\neg\neg\sigma$ has been deduced from Σ then σ may be deduced from Σ in one further step.
- (Contradiction Rule) If σ and $\neg \sigma$ have been deduced from Σ then $\bot$ may be deduced from Σ in one further step.
- (Reductio Ad Absurdum Rule) If $\bot$ has been deduced from $\Sigma \cup \{\sigma\}$ then $\neg \sigma$ may be deduced from Σ in one further step.

The notation $\Sigma \vdash \sigma$ will be used as in previous chapters to mean that σ may be deduced from Σ according to the rules above. Here, some set of propositional letters X is understood and $\Sigma \subseteq \mathrm{BT}(X)$, $\sigma \in \mathrm{BT}(X)$. So $\Sigma \vdash \sigma$ means there is a formal proof, i.e. a finite structured list of statements and subproofs of $\mathrm{BT}(X)$ obeying the above rules, showing that σ can be deduced from assumptions from Σ. In particular we will say that Σ is *inconsistent* if $\Sigma \vdash \bot$, and *consistent* otherwise.

We shall give some examples of formal proofs here, and in so doing we show how formal proofs are usually written, and spell out in more detail how the above rules should be applied.

Example 6.4 Let $X = \{a, b\}$ and $\Sigma = \{(\neg a \vee b)\}$. Then $\Sigma \vdash \neg(a \wedge \neg b)$.

Formal proof

$(a \wedge \neg b)$		(1)	Assumption
a		(2)	$\wedge$-Elimination
	$\neg a$	(3)	Assumption
	$\bot$	(4)	Contradiction
$\neg \neg a$		(5)	RAA
$\neg b$		(6)	$\wedge$-Elimination
$(\neg a \vee b)$		(7)	Given, from Σ
b		(8)	$\vee$-Elimination
$\bot$		(9)	Contradiction
$\neg(a \wedge \neg b)$		(10)	RAA

Example 6.5 Let $X = \{a, b\}$. Then $\varnothing \vdash (((a \wedge \neg b) \vee \neg a) \vee b)$.

Formal proof

$\neg(((a \wedge \neg b) \vee \neg a) \vee b)$		(1)	Assumption
	b	(2)	Assumption
	$(((a \wedge \neg b) \vee \neg a) \vee b)$	(3)	$\vee$-Introduction
	$\bot$	(4)	Contradiction
$\neg b$		(5)	RAA
	$\neg a$	(6)	Assumption
	$((a \wedge \neg b) \vee \neg a)$	(7)	$\vee$-Introduction
	$(((a \wedge \neg b) \vee \neg a) \vee b)$	(8)	$\vee$-Introduction
	$\bot$	(9)	Contradiction
$\neg \neg a$		(10)	RAA
a		(11)	$\neg$-Elimination
$(a \wedge \neg b)$		(12)	$\wedge$-Introduction
$((a \wedge \neg b) \vee \neg a)$		(13)	$\vee$-Introduction
$(((a \wedge \neg b) \vee \neg a) \vee b)$		(14)	$\vee$-Introduction
$\bot$		(15)	Contradiction
$\neg \neg(((a \wedge \neg b) \vee \neg a) \vee b)$		(16)	RAA
$(((a \wedge \neg b) \vee \neg a) \vee b)$		(17)	$\neg$-Elimination

This system is considerably more complicated than systems previously presented in this book. It does, however, share many features with other systems. In particular, the derived objects in a proof are simply strings of symbols with no *a priori* meanings attached. (Though there is a clear idea that such 'meaning' should be connected with boolean algebras – this will be taken up in the next chapter.) The proof rules are more complicated and there are more of them. However, they are not so difficult to learn, especially when you realise they all come in pairs, introduction and elimination, for each main symbol. The rules, if you read them carefully, should all be formalised versions of deduction rules that you are already familiar with and probably use in your own mathematical arguments anyway.

The proof rules, as in all the systems of this book, are *checkable*, meaning that it should be a mechanical process to verify whether a proposed proof is indeed a proof obeying the precise rules given. Each rule (apart from the Top Rule and the Given Statements Rule) have a number of 'inputs', usually one or two other previously deduced statements, but in the case of Reductio Ad Absurdum this 'input' is a 'subproof' of $\perp$ from an assumption σ.

If we were being stricter about our system we might insist that the line numbers of these previous 'inputs', the assumptions required for each rule, are indicated clearly in the narrative to the right of the deduction, and we will occasionally do this for emphasis. In practice, there is rarely any problem identifying these 'inputs', as the examples we will look at are short enough, though further identification certainly does not do any harm. It is worth remembering that every incorrect proof will have one or more lines that can be identified as not obeying the rules.

It is important, however, to be clear about exactly *which* previous lines in a proof are *available* as such 'inputs' for the next proof step, since proofs in this propositional logic typically involve several nested Reductio Ad Absurdum subproofs. This is indicated in the notation we use by the indentation level or vertical lines, and it is well worth drawing these lines properly when devising a proof. The rule is that an instance of a statement σ is available as an input to a deduction step S if: (a) it is above S on the page; *and* (b) it is in the same subproof or a direct parent-subproof. In terms of vertical lines, rule (b) means that none of the vertical lines to the left of the input statement σ are allowed to have completed before the step S being considered.

Note also that *any statement τ whatsoever* can be the assumption at the beginning of a subproof block. However, that subproof must end (at the same indentation level) with the statement $\perp$ for Reductio Ad Absurdum to apply; after this the subproof is closed, we are at a lower indentation level, and all statements in the subproof just closed (including the assumption) are no longer available as 'inputs' to further deductions.

Remark 6.6 The terminology sometimes used is that the assumption τ that leads to $\perp$ is *discharged* by the Reductio Ad Absurdum rule, and this is intended to indicate that the assumption need not be part of any 'global' assumptions that must be stated on the left hand side of the turnstile symbol, $\vdash$.

We follow with some instructive examples of incorrect proofs.

Example 6.7 Consider the following erroneous proof the shows b is a consequence of $(a \lor b)$. Of course, no such proof should be possible.

Formal proof

$(a \lor b)$	(1)	Given
$\neg b$	(2)	Assumption
$\neg a$	(3)	Assumption
b	(4)	$\lor$-Elimination, lines 1 and 3
$\bot$	(5)	Contradiction, lines 2 and 4
$\neg \neg b$	(6)	RAA
b	(7)	$\neg$-Elimination

Here, the error is that we do not keep track of the assumptions. The first assumption $\neg b$ starts a subproof and is done correctly, but the second assumption $\neg a$ should start an inner proof block, and this assumption is not discharged properly. When this is corrected, the proof looks like,

Formal proof

$(a \lor b)$	(1)	Given
$\neg b$	(2)	Assumption
$\neg a$	(3)	Assumption
b	(4)	$\lor$-Elimination, lines 1 and 3
$\bot$	(5)	Contradiction, lines 2 and 4
$\neg \neg a$	(6)	RAA
$\ldots$	(7)	

and, although we are not in a position to prove this here yet, there is no way to discharge the assumption $\neg b$.

Example 6.8 The following incorrect formal derivation shows how it is necessary to be extra careful with the positioning of brackets.

Formal proof

$a \lor b$	(1)	Given
$\neg a$	(2)	Assumption
$\neg a \lor b$	(3)	$\lor$-Introduction
$\bot$	(4)	Contradiction, lines 1 and 3
$\neg \neg a$	(5)	RAA
a	(6)	$\neg$-Elimination

In this example, brackets were carelessly omitted around the assumption $(a \vee b)$, and $(\neg a \vee b)$ in line 3, resulting in an unwanted application of the contradiction rule. (Line 3 is intended to mean $(\neg a) \vee b$, though brackets round a negation like this are not necessary – or even correct – according to the definition of boolean terms.)

Strictly speaking, brackets should always be included, and this would be the preferred option for a truly mechanical implementation of the proof rules, on a computer perhaps. But this quickly gets tedious, so we normally omit them according to some conventions. So even if you prefer to omit 'unnecessary' brackets, a little care and common sense should ensure that you do the Right Thing. With practice and experience, the sorts of errors illustrated above go against common sense and require a deliberate misinterpretation of the rules.

A more careful treatment of propositional logic than we are giving here would include the Unique Readability Theorem that says each boolean term can only be read in one way – unlike in our example where $\neg a \vee b$ was read as $(\neg a) \vee b$ and $\neg(a \vee b)$. The statement and proof of Unique Readability can be found on the companion web-pages.

By and large, in this book I shall take a relaxed view to notation, preferring to omit brackets where it is clear to do so and using standard mathematical notation where possible. My preference is to say that the $\neg$ operation binds more tightly than $\vee$ and $\wedge$, so $\neg a \vee b$ is always read as $(\neg a) \vee b$ and not $\neg(a \vee b)$. On the other hand, unlike some authors, I prefer not to give a distinction between $\vee$ and $\wedge$, for example always distinguishing $a \vee b \wedge c$ using brackets as $(a \vee b) \wedge c$ or $a \vee (b \wedge c)$. Terms such as $a \vee b \vee c \vee d$ will be assumed to associate to the left as $(((a \vee b) \vee c) \vee d)$, and similarly for $\wedge$.

Here is a correct proof, using the rules exactly as stated.

Example 6.9 $\{(a \wedge b)\} \vdash \neg(\neg a \vee \neg b)$

Formal proof

$(a \wedge b)$	(1)	Given
a	(2)	$\wedge$-Elimination
b	(3)	$\wedge$-Elimination
$\quad (\neg a \vee \neg b)$	(4)	Assumption
$\quad\quad \neg a$	(5)	Assumption
$\quad\quad \perp$	(6)	Contradiction, lines 2 and 5
$\quad \neg\neg a$	(7)	RAA
$\quad \neg b$	(8)	$\vee$-Elimination, lines 4 and 7
$\quad \perp$	(9)	Contradiction, lines 3 and 8
$\neg(\neg a \vee \neg b)$	(10)	RAA

As with other systems in this book, the proof rules should be used in a mechanical fashion without any attempt to interpret the symbols with 'meanings'. Thus, for example, a and $(\neg a \vee \neg b)$ do not together give $\neg b$ directly by $\vee$-Elimination. The above example shows a way this problem can be circumvented.

Manoeuvres such as 'from α deduce $\neg\neg\alpha$' and 'from α and $(\neg\alpha \vee \neg\beta)$ deduce $\neg\beta$' are sufficiently common that they are worth learning as *metarules* which can be safely used (when it is explained what you are doing) when the proof rules are used in a less rigid fashion. See Exercise 6.20 and the discussion following for more information on metarules.

Another place where we relax notation is in the notation on the left hand side of a turnstile symbol. Instead of using set theory notation with $\{\ldots\}$, $\cup$, $\varnothing$, etc., it is traditional to list formulas and sets of formulas, separating them with commas, and regard the list as a single *set* of formulas, so the order of formulas in the list and any repetitions in it is ignored. This applies to both the $\vdash$ turnstile of this chapter and the $\vDash$ turnstile that will be introduced in the next. Thus, with all the conventions in place, the previous example would be written as $a \wedge b \vdash \neg(\neg a \vee \neg b)$. The empty set is written as an empty list, as in $\vdash (a \vee \neg a)$.

To many students, writing informal mathematical proofs accurately is a sufficiently daunting prospect. It might seem that writing formal proofs following precise rules is much worse. In fact, I believe that the opposite is the case, and the formal rules add structure and suggest specific and useful proof strategies that not only make writing formal proofs easier, but can help in suggesting ways to write informal arguments too.

When constructing a proof, the first step is to write down all the assumptions or 'known data' at the top of a page and, leaving a large space, write the 'goal' or statement-to-be-proved at the bottom. Then a small number of memorable 'proof strategies' can be employed to generate most of the structure of the proof and (for propositional logic, at least) these can in fact generate the whole proof with little or no real thinking required at all.

Many of the strategies are common sense from the rules. For example, one is that, having got the statement $\alpha \wedge \beta$ it makes sense to deduce immediately α and β by $\wedge$-Elimination. This was done for example in lines 2 and 3 of the previous example.

Another strategy is that, to prove a statement $\neg\alpha$ it makes sense to start a new subproof with assumption α and try and to $\bot$. Our required $\neg\alpha$ then follows from Reductio Ad Absurdum. This strategy was used on line 4 of the previous example.

A variation of this strategy is also useful, but perhaps in a slightly more limited way: to prove a statement α it is always possible to start a new subproof with assumption $\neg \alpha$ and try to prove $\bot$. Our required α then follows from Reductio Ad Absurdum and $\neg$-Elimination. You will find plenty of occasions where this strategy has been used. It is often essential, and even when it is not essential it can never do any harm, though it sometimes results in longer or less elegant proofs, so should probably be left as a last resort. It is also worth mentioning that any statement at all can be derived from a contradiction, as will be shown in Example 6.10 below.

Of course, if applying a strategy such as one of these results in a new statement being assumed or known 'true' in a part of a proof, then other strategies should be applied to that new statement, and so on until (hopefully) the proof is complete.

The next few examples give a short but complete list of strategies for the propositional logic with connectives $\top, \bot, \neg, \wedge, \vee$.

Example 6.10 From $\bot$ or $\neg \top$ any statement can be deduced.

Formal proof

$\bot$	(1)	Given
$\quad \neg \theta$	(2)	Assumption
$\quad \bot$	(3)	Line 1
$\neg \neg \theta$	(4)	RAA
θ	(5)	$\neg$-Elimination

It is part of the rules that statements can be copied from previously deduced ones. In line 3 the statement $\bot$ is available from line 1, so a contradiction is immediate. If you prefer, think of line 3 as repetition or 're-deduction' of $\bot$ from whatever was used to get $\bot$ in line 1.

A related formal proof shows that from $\neg \top$ we quickly get $\bot$. This is as follows.

Formal proof

$\neg \top$	(1)	Given
$\top$	(2)	Top rule
$\bot$	(3)	Contradiction, lines 1 and 2

Example 6.11 To prove the statement θ from $(\alpha \wedge \beta)$ and other given statements Σ, deduce both of α and β first by $\wedge$-Elimination and prove θ from Σ, α and β.

Formal proof

$(\alpha \wedge \beta)$	(1)	Given
α	(2)	$\wedge$-Elimination
β	(3)	$\wedge$-Elimination
...	(4)	
θ	(5)	

Example 6.12 To prove a statement θ from $\neg(\alpha \vee \beta)$ and other given statements Σ, deduce both of $\neg \alpha$ and $\neg \beta$ first, as shown in the following argument, and then prove θ from Σ, $\neg \alpha$ and $\neg \beta$.

Formal proof

$\neg(\alpha \vee \beta)$	(1)	Given
$\quad \alpha$	(2)	Assumption
$\quad (\alpha \vee \beta)$	(3)	$\vee$-Introduction
$\quad \perp$	(4)	Contradiction
$\neg \alpha$	(5)	RAA

The proof that $\neg(\alpha \vee \beta) \vdash \neg \beta$ is similar.

It is not so easy to see how to use a statement of the form $(\alpha \vee \beta)$ as an 'input', since we are not told which of α and β is true. The solution in such cases is to construct proofs from *both* α and β.

Example 6.13 To prove a statement θ from $(\alpha \vee \beta)$ and other given statements Σ, try to construct *two* proofs of θ, one from α and Σ, and the other from β and Σ. The final argument can be put together like this.

Formal proof

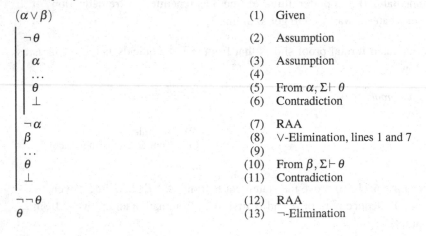

$(\alpha \vee \beta)$	(1)	Given
$\quad \neg \theta$	(2)	Assumption
$\quad \quad \alpha$	(3)	Assumption
$\quad \quad ...$	(4)	
$\quad \quad \theta$	(5)	From $\alpha, \Sigma \vdash \theta$
$\quad \quad \perp$	(6)	Contradiction
$\quad \neg \alpha$	(7)	RAA
$\quad \beta$	(8)	$\vee$-Elimination, lines 1 and 7
$\quad ...$	(9)	
$\quad \theta$	(10)	From $\beta, \Sigma \vdash \theta$
$\quad \perp$	(11)	Contradiction
$\neg \neg \theta$	(12)	RAA
θ	(13)	$\neg$-Elimination

In the special case when θ is $\perp$, the previous proof simplifies to the following formal proof.

Formal proof

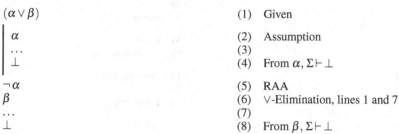

$(\alpha \lor \beta)$	(1)	Given
α	(2)	Assumption
...	(3)	
$\perp$	(4)	From $\alpha, \Sigma \vdash \perp$
$\neg \alpha$	(5)	RAA
β	(6)	$\lor$-Elimination, lines 1 and 7
...	(7)	
$\perp$	(8)	From $\beta, \Sigma \vdash \perp$

Using a given statement of the form $\neg(\alpha \land \beta)$ is similar to the case using a statement $(\neg \alpha \lor \neg \beta)$, and you should construct two proofs and put them together using an assumption and subproof.

Example 6.14 To prove a statement θ from $\neg(\alpha \land \beta)$ and other statements Σ, construct two proofs of θ, one from $\neg \alpha$ and Σ, and the other from $\neg \beta$ and Σ.

Formal proof

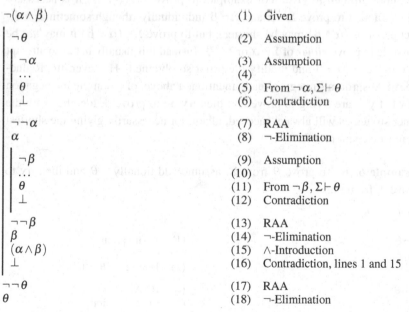

$\neg(\alpha \land \beta)$	(1)	Given
$\neg \theta$	(2)	Assumption
$\neg \alpha$	(3)	Assumption
...	(4)	
θ	(5)	From $\neg \alpha, \Sigma \vdash \theta$
$\perp$	(6)	Contradiction
$\neg \neg \alpha$	(7)	RAA
α	(8)	$\neg$-Elimination
$\neg \beta$	(9)	Assumption
...	(10)	
θ	(11)	From $\neg \beta, \Sigma \vdash \theta$
$\perp$	(12)	Contradiction
$\neg \neg \beta$	(13)	RAA
β	(14)	$\neg$-Elimination
$(\alpha \land \beta)$	(15)	$\land$-Introduction
$\perp$	(16)	Contradiction, lines 1 and 15
$\neg \neg \theta$	(17)	RAA
θ	(18)	$\neg$-Elimination

In the special case when θ is $\perp$, this simplifies to the following formal proof.

Formal proof

$\neg(\alpha \wedge \beta)$	(1)	Given
$\quad \neg\alpha$	(2)	Assumption
$\quad \ldots$	(3)	
$\quad \bot$	(4)	From $\neg\alpha, \Sigma \vdash \bot$
$\neg\neg\alpha$	(5)	RAA
α	(6)	$\neg$-Elimination
$\quad \neg\beta$	(7)	Assumption
$\quad \ldots$	(8)	
$\quad \bot$	(9)	From $\neg\beta, \Sigma \vdash \bot$
$\neg\neg\beta$	(10)	RAA
β	(11)	$\neg$-Elimination
$(\alpha \wedge \beta)$	(12)	$\wedge$-Introduction
$\bot$	(13)	Contradiction, lines 1 and 15

Example 6.15 To prove θ from $\neg\neg\alpha$, first deduce α from $\neg$-Elimination and then try to prove θ from this.

This forms a complete set of strategies on the 'assumptions' side for propositional proofs using $\bot, \neg, \wedge, \vee$. The *best* strategies on the *conclusion* side are more difficult to give. For example, to prove $\Sigma \vdash (\alpha \wedge \beta)$ it is necessary and sufficient to prove $\Sigma \vdash \alpha$ and $\Sigma \vdash \beta$ individually (though sometimes a direct proof of $(\alpha \wedge \beta)$ may be shorter), but to prove $\Sigma \vdash (\alpha \vee \beta)$ it may *not* be possible to prove *either* of $\Sigma \vdash \alpha$ or $\Sigma \vdash \beta$. Instead it is usually better to attempt to prove $\Sigma, \neg\alpha \vdash \beta$ and modify the proof so obtained. However, the Reductio Ad Absurdum strategy, already mentioned above, of assuming the negation of what you are trying to prove and then trying to prove $\bot$ together with the other strategies will always succeed, albeit not necessarily giving the shortest or most elegant proofs.

Example 6.16 To prove θ from Σ, assume additionally $\neg\theta$ and then try to prove $\bot$ from this.

Formal proof

$\quad \neg\theta$	(1)	Assumption
$\quad \ldots$	(2)	
$\quad \bot$	(3)	From $\Sigma, \neg\theta \vdash \bot$
$\neg\neg\theta$	(4)	RAA
θ	(5)	$\neg$-Elimination

The following section gives further examples, including discussion of the implication operator, and the final optional section of this chapter sketches a proof that the above examples do indeed form a complete set of strategies that are guaranteed to produce a proof of an assertion in propositional logic – provided, of course, that such a proof exists at all.

6.2 Examples and exercises

Exercise 6.17 Give a formal proof of the *law of excluded middle* which says that if ϕ is any statement or boolean term then $(\phi \vee \neg \phi)$ has a formal proof in the system above.

Exercise 6.18 Find a formal proof of $(\neg a \vee \neg b)$ from $\neg(a \wedge b)$, and another formal proof for the converse.

Exercise 6.19 Give formal derivations showing the following:

- $\neg p \vee q \vdash \neg \neg \neg q \vee \neg p$
- $\vdash \neg (\neg p \vee \bot) \vee \neg p$
- $\neg p \vee q, \neg p \vee r \vdash \neg p \vee (q \wedge r)$
- $\vdash \neg p \vee (p \vee q)$
- $\neg p \vee q \vdash \neg (\neg q \vee r) \vee (\neg p \vee r)$

Exercise 6.20 Show that the following rule could be added to the system without changing the set of statements that are provable: 'If τ has been deduced from $\Sigma \cup \{\neg \sigma\}$, then $(\sigma \vee \tau)$ may be deduced from Σ in one further step.' (Hint: show how to transform a proof of τ from $\Sigma \cup \{\neg \sigma\}$ to one of $(\sigma \vee \tau)$ from Σ.)

The last exercise shows how 'metarules' can be added to the system; such rules can simplify the process of finding formal proofs considerably, as well as shortening and making the resulting proofs more readable.

As an example of metarules, note that the $\vee$-elimination rule is quite tricky as it is only available in one limited form, and like all the other rules it is a syntactic rule and only applies when the symbols combine in exactly the way stated in the rule. Fortunately, all the other forms that most people expect to be present can be justified as metarules.

Exercise 6.21 Show that addition of the following metarules to the system does not change the set of sentences provable.

- 'If σ and $(\neg \sigma \vee \tau)$ have been deduced from Σ then τ may be deduced from Σ in one further step.'
- 'If $\neg \tau$ and $(\sigma \vee \tau)$ have been deduced from Σ then σ may be deduced from Σ in one further step.'
- 'If τ and $(\sigma \vee \neg \tau)$ have been deduced from Σ then σ may be deduced from Σ in one further step.'

Exercise 6.22 Add a new boolean connective $\rightarrow$ for 'implication' together with two additional rules representing the way mathematicians tend to use this word.

- ($\rightarrow$-Introduction) If τ can be deduced from $\Sigma \cup \{\sigma\}$ then $(\sigma \rightarrow \tau)$ can be deduced from Σ in one further step.
- ($\rightarrow$-Elimination, also called *modus ponens*) If $(\sigma \rightarrow \tau)$ and σ can be deduced from Σ then τ can be deduced from Σ in one further step.

Show that $(\sigma \rightarrow \tau)$ and $(\neg \sigma \vee \tau)$ are equivalent by proving one statement from the other in the new system.

Some differences between the 'natural language' implication and mathematical implication have already been discussed. Another difference, at least for some speakers, is that in natural language '*a* implies *b*' is typically interpreted as implying that *a* could hold or is at least *possible* in some hypothetical situation. One consequence is that, in natural language '*a* implies *b*' and '*a* implies not-*b*' seem contradictory, for in a hypothetical world where *a* holds it cannot be the case that both *b* and not-*b* hold. This was the source of controversy between two nineteenth century Oxford logicians, Dodgson (aka Lewis Carroll) and Cook Wilson. Dodgson invented a story, known as the *Barber Shop Paradox*, to illustrate the difficulties that the natural language notion of implication has when combined with Reductio Ad Absurdum, and it seems he was surprised when Cook Wilson argued back [3]. The essence of the 'paradox' was that assuming $\neg a \rightarrow b$ and $\neg a \rightarrow \neg b$ are contradictory it is possible to prove *c* from the statements $a \rightarrow b$, $\neg a \rightarrow \neg b$, and $a \vee b \vee c$. (Assume $\neg c$ and observe that $a \vee b \vee c$ then gives $\neg a \rightarrow b$.) Of course this can be cleared up very quickly with the formal propositional logic in this chapter (something that was not available in Oxford in Dodgson and Cook Wilson's day).

Exercise 6.23 Show that:

(a) $(a \rightarrow b) \wedge (a \rightarrow \neg b)$ and $\neg a$ are equivalent in propositional logic;

(b) $a \rightarrow b,\ \neg a \rightarrow \neg b,\ a \vee b \vee c \vdash \neg c \rightarrow a$;

(c) $a \rightarrow b,\ \neg a \rightarrow \neg b,\ a \vee b \vee c \nvdash c$.

(Part (c) is possibly too hard at this stage, but will become easy with the results of the next chapter.)

We have seen many kinds of implication in this book, including the informal use of 'implies' in our metatheorems and metaproofs, $\vDash$, $\vdash$, and now $\rightarrow$. They are all different (albeit, related) and it is important to be able to distinguish between them all. The next exercises point out some connections between $\vdash$ and $\rightarrow$.

Exercise 6.24 Prove the *Deduction Theorem* that says that $\Sigma \cup \{\sigma\} \vdash \tau$ is true if and only if $\Sigma \vdash \sigma \rightarrow \tau$ is true, where $\vdash$ is for the extended system with the additional rules for $\rightarrow$ added. (The Deduction Theorem is quite straightforward for the system we have here given the rules for $\rightarrow$, but for other formal systems becomes an important result.)

Exercise 6.25 Show that (a) $\neg \theta \vdash \neg \psi$ implies $\psi \vdash \theta$, and (b) in the propositional logic with $\rightarrow$ added we have

$$\vdash ((\neg \theta \rightarrow \neg \psi) \rightarrow (\psi \rightarrow \theta)).$$

In what ways are statements (a) and (b) different?

6.3 Decidability of propositional logic*

Given a finite set of boolean terms Σ and a further boolean term σ we may ask whether $\Sigma \vdash \sigma$. More generally, the data Σ, σ being finite could be the input to a computer program, and we could ask whether there is an algorithm, programmable on a computer, that correctly tells us whether $\Sigma \vdash \sigma$, and, if possible, generates a formal proof too.

For propositional logic, this is indeed possible. One method uses valuations, as discussed in the next chapter, but this method simply yields a yes/no answer, not a formal proof. Another method uses the proof strategies discussed informally in the text above.

The idea of the algorithm is to reduce the question 'does $\Sigma \vdash \sigma$?' to 'simpler' questions, and working backwards each stage of the reduction gives a recipe of how to transform proofs of the 'simpler' questions to one of the original. More precisely, the following theorem (which is a formal statement of the idea of proof strategies discussed above) allows us to reduce $\Sigma \vdash \sigma$ to a number of questions about provability of statements with no $\wedge$, $\vee$, or double negation symbols.

Theorem 6.26 *Let X be a set of propositional letters, Σ a finite set of boolean terms in X, and α, β, σ other boolean terms in X. Then:*

(i) $\Sigma \vdash \sigma$ *if and only if* $\Sigma, \neg\sigma \vdash \bot$;

(ii) $\Sigma, \neg\neg\alpha \vdash \bot$ *if and only if* $\Sigma, \alpha \vdash \bot$;

(iii) $\Sigma, \alpha \wedge \beta \vdash \bot$ *if and only if* $\Sigma, \alpha, \beta \vdash \bot$;

(iv) $\Sigma, \neg(\alpha \vee \beta) \vdash \bot$ *if and only if* $\Sigma, \neg\alpha, \neg\beta \vdash \bot$;

(v) $\Sigma, \alpha \vee \beta \vdash \bot$ *if and only if* $\Sigma, \alpha \vdash \bot$ *and* $\Sigma, \beta \vdash \bot$;

(vi) $\Sigma, \neg(\alpha \wedge \beta) \vdash \bot$ *if and only if* $\Sigma, \neg\alpha \vdash \bot$ *and* $\Sigma, \neg\beta \vdash \bot$.

Proof The right-to-left directions were covered in examples earlier in this chapter. The other directions are done in a similar way.

(i) If $\Sigma \vdash \sigma$ then $\Sigma, \neg\sigma \vdash \bot$ as

Formal proof

$\neg\sigma$	(1)	Given
$\ldots$	(2)	
σ	(3)	From $\Sigma \vdash \sigma$
$\bot$	(4)	Contradiction, lines 1 and 3

(ii) If $\Sigma, \neg\neg\alpha \vdash \bot$ then $\Sigma, \alpha \vdash \bot$ as

Formal proof

α	(1)	Given
$\quad\neg\alpha$	(2)	Assumption
$\quad\bot$	(3)	Contradiction, lines 1 and 2
$\neg\neg\alpha$	(4)	RAA
$\ldots$	(5)	
$\bot$	(6)	From $\Sigma, \neg\neg\alpha \vdash \bot$

The other cases are similarly easy, and we leave them as exercises.   □

By iterating the previous theorem, we see that any statement of the form $\Sigma \vdash \sigma$ is equivalent to a number of statements of the form $\Pi \vdash \bot$, where each $\alpha \in \Pi$ is either $\top$, $\bot$, $\neg\top$, $\neg\bot$, or a or $\neg a$ for some propositional letter a. What is more, the proof of the theorem shows how we can transform formal proofs of all the statements $\Pi \vdash \bot$ back into a proof of the original $\Sigma \vdash \sigma$.

Statements of the form $\top$, $\bot$, or a are called *atomic*. Those of the form $\neg\top$, $\neg\bot$, or $\neg a$ are called *negated atomic*. This means that we can reduce our problem of whether $\Sigma \vdash \sigma$ to the case where $\sigma = \bot$ and Σ consists of only atomic or negated atomic statements. Now two cases when such a set Σ is obviously contradictory are (a) when Σ contains $\bot$ or $\neg\top$, and (b) when Σ

contains both a and $\neg a$ for some letter a. In fact, this is the only way such a set Σ can prove $\bot$, as the following shows.

Theorem 6.27 *Let Σ consist of atomic and negated atomic statements only. Then $\Sigma \vdash \bot$ if and only if Σ contains $\bot$ or $\neg\top$, or both a and $\neg a$ for some letter a.*

I am not going to prove this here. The difficult direction is to show that if Σ does not contain $\bot$ nor $\neg\top$, nor both a and $\neg a$ for some a, then $\Sigma \nvdash \bot$, and the worry is that somehow the rules for $\vee$ or $\wedge$ might be used in a way that we have not yet envisaged. One method (that will be discussed in more detail in the next chapter) is to define a semantics making every statement a occurring in Σ 'true' and every other letter b 'false' and use this as a basis for induction on the length of proofs. Another method, also by induction on the length of proofs, is to show that if such $\Sigma \vdash \bot$ then there is a proof that does not use any of the rules for $\vee$ or $\wedge$.

Either way, our discussion yields the following theorem.

Theorem 6.28 *There is an algorithm which, on input Σ and σ (a finite set of statements and a single statement in the propositional logic) correctly determines whether $\Sigma \vdash \sigma$, and in case the answer is 'yes' also returns a formal proof of σ from Σ.*

7

Valuations

7.1 Semantics for propositional logic

Following the general method for other formal systems in this book, we must connect the system for propositional logic of the last chapter with the boolean algebras of the chapter preceding it, by using boolean algebras to provide 'meanings' or *semantics* for boolean terms and the symbolic manipulations in the system for proof. As for other logics in this book, we will develop our semantics far enough to present a Completeness Theorem and a Soundness Theorem. The basis of our semantics is the following very simple idea of a valuation.

Definition 7.1 Let X be any set, and B a boolean algebra. A *valuation* on X is a function $f: X \to B$.

This notion of a valuation is very straightforward, but is enough to give a valuation $v: X \to B$ interesting extra structure. A valuation induces a map $BT(X) \to B$ defined by evaluating boolean terms over X in the boolean algebra B, using the value $v(x)$ in place of the symbol x from X and the operations in B for all other terms. Formally, this means making the following definition by induction on the number of symbols in a term:

- $v(\neg \sigma) = (v(\sigma))'$
- $v(\sigma \wedge \tau) = (v(\sigma)) \wedge (v(\tau))$
- $v(\sigma \vee \tau) = (v(\sigma)) \vee (v(\tau))$
- $v(\top) = \top$
- $v(\bot) = \bot$

where the right hand side of each of these is evaluated in B using its boolean algebra structure. Remember that the arguments inside the $v(\ldots)$ are simply formal terms, i.e. strings of symbols in the appropriate alphabet, so the symbols on the right and left hand sides of the above equations are interpreted quite differently, in one case as 'pure symbols' in boolean terms, in the other as operations in B.

Since $BT(X) \supseteq X$, this valuation function $BT(X) \to B$ extends v, and it will not cause any confusion to denote it with the same letter, also writing it as $v: BT(X) \to B$.

Many valuations use the simplest possible boolean algebra $2 = \{\top, \bot\}$ for the target boolean algebra B. To specify such a valuation it suffices to give a value ($\top$, meaning true, or $\bot$, meaning false) to each element of X. Indeed, one of the results we shall prove about propositional logic later on is that this two-element boolean algebra suffices to give an adequately rich semantics for the proof system. But at the outset, this is not obvious.

Example 7.2 If $X = \{a, b, c\}$ then we have the following.

(i) All valuations $v: X \to 2$ make the boolean terms $(((a \wedge \neg b) \vee \neg a) \vee b)$, $(a \vee \neg a)$, and $\neg \bot$ true ($\top$).

(ii) The boolean terms $\neg \top$, $(c \wedge \neg c)$ and $(a \wedge (b \wedge (\neg a \vee \neg b)))$ are terms for which every valuation $v: X \to 2$ makes them false ($\bot$).

(iii) Terms such as $((a \wedge \neg b) \vee c)$ and $((\neg c \vee \neg b) \wedge (c \vee \neg a))$ are sometimes true, sometimes false, depending on the choice of valuation.

To see this, one needs to go carefully through all of the possible combinations of values from $\top, \bot$ that can be given to the letters a, b, c. The best way to set this out is in a table (called a *truth table*). Each row of the truth table corresponds to a possible combination of values for the proposition letters, and the columns of the truth table correspond to expressions or subexpressions that have to be evaluated in the boolean algebra $2 = \{\top, \bot\}$.

For the first,

a	b	$\neg b$	$(a \wedge \neg b)$	$\neg a$	$((a \wedge \neg b) \vee \neg a)$	$(((a \wedge \neg b) \vee \neg a) \vee b)$
$\top$	$\top$	$\bot$	$\bot$	$\bot$	$\bot$	$\top$
$\top$	$\bot$	$\top$	$\top$	$\bot$	$\top$	$\top$
$\bot$	$\top$	$\bot$	$\bot$	$\top$	$\top$	$\top$
$\bot$	$\bot$	$\top$	$\bot$	$\top$	$\top$	$\top$

This justifies the assertion that $v(((a \wedge \neg b) \vee \neg a) \vee b) = \top$ for all such valuations.

Now consider:

a	b	$\neg a$	$\neg b$	$(\neg a \vee \neg b)$	$(b \wedge (\neg a \vee \neg b))$	$(a \wedge (b \wedge (\neg a \vee \neg b)))$
$\top$	$\top$	$\bot$	$\bot$	$\bot$	$\bot$	$\bot$
$\top$	$\bot$	$\bot$	$\top$	$\top$	$\bot$	$\bot$
$\bot$	$\top$	$\top$	$\bot$	$\top$	$\top$	$\bot$
$\bot$	$\bot$	$\top$	$\top$	$\top$	$\bot$	$\bot$

So $v(a \wedge (b \wedge (\neg a \vee \neg b)))$ is $\bot$ for all valuations.

Finally,

a	b	c	$\neg b$	$(a \wedge \neg b)$	$((a \wedge \neg b) \vee c)$
$\top$	$\top$	$\top$	$\bot$	$\bot$	$\top$
$\top$	$\top$	$\bot$	$\bot$	$\bot$	$\bot$
$\top$	$\bot$	$\top$	$\top$	$\top$	$\top$
$\top$	$\bot$	$\bot$	$\top$	$\top$	$\top$
$\bot$	$\top$	$\top$	$\bot$	$\bot$	$\top$
$\bot$	$\top$	$\bot$	$\bot$	$\bot$	$\bot$
$\bot$	$\bot$	$\top$	$\top$	$\bot$	$\top$
$\bot$	$\bot$	$\bot$	$\top$	$\bot$	$\bot$

Thus $((a \wedge \neg b) \vee c)$ has value $\top$ for some valuations and $\bot$ for others. We leave the reader to check the remaining examples.

Note that in cases such as $((\neg c \vee \neg b) \wedge (c \vee \neg a))$ with three proposition letters, there are $8 = 2^3$ combinations, as there are 2 truth values and 3 letters, so the truth table has 8 rows, not counting the header row. In general a truth table with n proposition letters will have 2^n rows.

Definition 7.3 A formula $\sigma \in \mathrm{BT}(X)$ such that each valuation $v : \mathrm{BT}(X) \rightarrow \{\top, \bot\}$ makes $v(\sigma) = \top$ is called a *tautology*. A formula $\sigma \in \mathrm{BT}(X)$ such that there is some valuation $v : \mathrm{BT}(X) \rightarrow \{\top, \bot\}$ such that $v(\sigma) = \top$ is called *satisfiable*.

Definition 7.4 Let $v : \mathrm{BT}(X) \rightarrow B$ be a valuation and let $\Sigma \subseteq \mathrm{BT}(X)$ be a set of boolean terms. Then $v(\Sigma)$ denotes the *set of values* that v takes in B on inputs from Σ, i.e. $v(\Sigma) = \{v(\sigma) : \sigma \in \Sigma\}$.

Definition 7.5 Let B be a boolean algebra and $A \subseteq B$ be a *finite* subset of B. Then, by repeated application of $\wedge$ and $\vee$, the set A has both a greatest lower bound and a least upper bound. These depend only on A and not the order in which $\wedge$ or $\vee$ was applied, and will be denoted $\bigwedge A$ and $\bigvee A$ respectively, using a larger symbol for the operator (analogous to the large union or intersection signs $\bigcup$, $\bigcap$ or the large summation and product signs $\sum$, $\prod$). To cover the case when A is empty, we conventionally define $\bigwedge \varnothing = \top$ and $\bigvee \varnothing = \bot$.

There are boolean algebras in which $\bigwedge A$ and $\bigvee A$ are defined for all sets A, not just the finite ones. The power set algebra $P(X)$ of a set X is an example. These are interesting algebras, but not every one is of this type. To ensure that our results apply to all boolean algebras, we will take care to apply $\bigwedge$ and $\bigvee$ to finite sets only.

Here is the formal definition of semantics, $\models$. It says that $\Sigma \models \sigma$ holds if σ is at least as true as the *greatest lower bound* (or 'and') of some finite subset of Σ. (We cannot take the whole of Σ, in case that Σ itself is infinite.) To measure 'at least as true' we use boolean algebras and valuations.

Definition 7.6 Given a set X, $\Sigma \subseteq \mathrm{BT}(X)$ and $\sigma \in \mathrm{BT}(X)$ we write $\Sigma \models \sigma$ for the statement that, for all boolean algebras B and all valuations $v: \mathrm{BT}(X) \to B$, there is a finite subset $\Sigma_0 \subseteq \Sigma$ such that $\bigwedge v(\Sigma_0) \leqslant v(\sigma)$.

Definition 7.7 We write $\Sigma \models_2 \sigma$ for the specialisation of $\Sigma \models \sigma$ to the boolean algebra $2 = \{\top, \bot\}$. That is, $\Sigma \models_2 \sigma$ means for all valuations $v: \mathrm{BT}(X) \to 2$, there is a finite subset $\Sigma_0 \subseteq \Sigma$ such that $\bigwedge v(\Sigma_0) \leqslant v(\sigma)$.

Once again, truth tables can help calculate whether $\Sigma \models_2 \sigma$ holds or not.

Example 7.8 Given proposition letters p, q we have

$$p, \neg(p \wedge r) \models_2 \neg((p \vee q) \wedge r).$$

The truth table is

p	q	r	p	$\neg(p \wedge r)$	$\neg((p \vee q) \wedge r)$
$\top$	$\top$	$\top$	$\top$	$\bot$	$\bot$
$\top$	$\top$	$\bot$	$\top$	$\top$	$\top$
$\top$	$\bot$	$\top$	$\top$	$\bot$	$\bot$
$\top$	$\bot$	$\bot$	$\top$	$\top$	$\top$
$\bot$	$\top$	$\top$	$\bot$	$\top$	$\bot$
$\bot$	$\top$	$\bot$	$\bot$	$\top$	$\top$
$\bot$	$\bot$	$\top$	$\bot$	$\top$	$\top$
$\bot$	$\bot$	$\bot$	$\bot$	$\top$	$\top$

Observe that in each of the two rows in which both p and $\neg(p \wedge r)$ are $\top$ (the second and fourth), the conclusion $\neg((p \vee q) \wedge r)$ is also $\top$. This suffices. The fact that the conclusion is also true in additional rows is irrelevant.

Example 7.9 Again, with proposition letters p, q we have

$$\neg(q \wedge r), (p \vee r) \not\vDash_2 p \wedge (\neg q \vee r).$$

The truth table is

p	q	r	$\neg(q \wedge r)$	$(p \vee r)$	$p \wedge (\neg q \vee r)$
$\top$	$\top$	$\top$	$\bot$	$\top$	$\top$
$\top$	$\top$	$\bot$	$\top$	$\top$	$\bot$
$\top$	$\bot$	$\top$	$\top$	$\top$	$\top$
$\top$	$\bot$	$\bot$	$\top$	$\top$	$\top$
$\bot$	$\top$	$\top$	$\bot$	$\top$	$\bot$
$\bot$	$\top$	$\bot$	$\top$	$\bot$	$\bot$
$\bot$	$\bot$	$\top$	$\top$	$\top$	$\bot$
$\bot$	$\bot$	$\bot$	$\top$	$\bot$	$\bot$

Here, we see that the valuation $v(p) = \bot$, $v(q) = \bot$, $v(r) = \top$ makes $\neg(q \wedge r)$ and $p \vee r$ both equal to $\top$ but makes the conclusion $p \wedge (\neg q \vee r)$ equal to $\bot$, so $\neg(q \wedge r), (p \vee r) \vDash_2 p \wedge (\neg q \vee r)$ is false.

Theorem 7.10 (Soundness Theorem) *Let X be a set and $\tau \in \mathrm{BT}(X)$, and suppose that $\Sigma \vdash \tau$. Then $\Sigma \vDash \tau$.*

Proof Let B be a boolean algebra and $v: \mathrm{BT}(X) \to B$ a valuation. Let Σ_0 be the set of assumptions used in Σ to prove τ. We shall prove $\bigwedge v(\Sigma_0) \leqslant v(\tau)$. To simplify notation we will assume without loss of generality that Σ is finite and $\Sigma_0 = \Sigma$.

So we must prove that if there is a formal derivation of τ from Σ then the value of the conclusion of such a formal derivation, $v(\tau)$, is always 'at least as true as' the value of the assumptions, Σ. In particular if $v(\sigma) = \top$ for all $\sigma \in \Sigma$ then $\top = \bigwedge v(\Sigma) \leqslant v(\tau)$, so $v(\tau) = \top$.

The proof is by induction on the length of a derivation. Our induction hypothesis $H(n)$ is the statement that if p is a derivation in the system with at most n steps and p is a proof of τ from a finite set of assumptions Σ, then $\bigwedge v(\Sigma) \leqslant v(\tau)$ for all valuations v on $\mathrm{BT}(X)$.

Assume we have such a derivation p of length n of τ from a finite set Σ, that $H(n-1)$ holds, and that v is a valuation. As before, we look at the very last deduction in p.

The cases when the last deduction used is either the Top Rule or the Given Statements Rule is easy, since in the first of these cases τ is $\top$ and $v(\top) = \top \geqslant \bigwedge v(\Sigma)$, and in the second of these cases τ is some $\sigma \in \Sigma$ and hence $v(\tau) = v(\sigma)$ for some $\sigma \in \Sigma$ so $v(\tau) \geqslant \bigwedge v(\Sigma)$.

If the last step in p is $\wedge$-Introduction then τ is $(\alpha \wedge \beta)$ where α, β have previously been derived from Σ. Thus $v(\alpha), v(\beta) \geqslant \bigwedge v(\Sigma)$ by the induction hypothesis and so $\bigwedge v(\Sigma)$ is a lower bound for both $v(\alpha), v(\beta)$, hence $v(\alpha \wedge \beta) = v(\alpha) \wedge v(\beta) \geqslant \bigwedge v(\Sigma)$.

If the last step in p is $\wedge$-Elimination then τ is α or β where $(\alpha \wedge \beta)$ has been previously derived from Σ. So by the induction hypothesis $v(\alpha \wedge \beta) \geqslant \bigwedge v(\Sigma)$. And as $v(\alpha \wedge \beta) = v(\alpha) \wedge v(\beta) \leqslant v(\alpha), v(\beta)$ we have $v(\alpha), v(\beta) \geqslant \bigwedge v(\Sigma)$ by transitivity.

If the last step in p is $\vee$-Elimination then $(\sigma \vee \tau)$ and $\neg \sigma$ have already been deduced from Σ so by the induction hypothesis $v(\sigma \vee \tau) = v(\sigma) \vee v(\tau) \geqslant \bigwedge v(\Sigma)$ and $v(\neg \sigma) = v(\sigma)' \geqslant \bigwedge v(\Sigma)$. So from $v(\sigma) \vee v(\tau) \geqslant \bigwedge v(\Sigma)$ we have

$$v(\sigma)' \wedge (v(\sigma) \vee v(\tau)) \geqslant v(\sigma)' \wedge \bigwedge v(\Sigma).$$

And from $v(\sigma)' \geqslant \bigwedge v(\Sigma)$ we have

$$v(\sigma)' \wedge \bigwedge v(\Sigma) = \bigwedge v(\Sigma).$$

Also

$$v(\sigma)' \wedge (v(\sigma) \vee v(\tau)) = v(\sigma') \wedge v(\tau)$$

by distributivity, so $v(\tau) \geqslant v(\sigma') \wedge v(\tau) \geqslant \bigwedge v(\Sigma)$, as required.

If the last step in p is $\neg$-Elimination then τ has already been derived so $v(\neg \neg \tau) = v(\tau)'' = v(\tau) \geqslant \bigwedge v(\Sigma)$, as required.

If the last step is the Contradiction Rule, both σ and $\neg \sigma$ have already been derived with $v(\sigma), v(\neg \sigma) \geqslant \bigwedge v(\Sigma)$. So $\bigwedge v(\Sigma)$ is a lower bound for $v(\sigma)$ and $v(\neg \sigma) = v(\sigma)'$ so $\perp = v(\sigma) \wedge v(\sigma)' \geqslant \bigwedge v(\Sigma)$, as required.

If the last step is the $\vee$-Introduction Rule, assume that τ is $(\alpha \vee \beta)$ or $(\beta \vee \alpha)$ where α has been deduced from Σ. Then by our induction hypothesis $v(\alpha) \geqslant \bigwedge v(\Sigma)$, so

$$v(\alpha \vee \beta), v(\beta \vee \alpha) \geqslant v(\alpha) \geqslant \bigwedge v(\Sigma)$$

as required.

Finally, the Reductio Ad Absurdum Rule: if $\perp$ has been deduced from $\Sigma \cup \{\alpha\}$ then by the induction hypothesis $\perp = v(\perp) \geqslant \bigwedge v(\Sigma) \wedge v(\alpha)$ so $v(\neg \alpha) = v(\alpha)' \vee v(\perp) \geqslant v(\alpha)' \vee (\bigwedge v(\Sigma) \wedge v(\alpha))$. But

$$v(\alpha)' \vee (\bigwedge v(\Sigma) \wedge v(\alpha)) = v(\alpha)' \vee \bigwedge v(\Sigma) \geqslant \bigwedge v(\Sigma)$$

by distributivity, so $v(\neg \alpha) = v(\alpha)' \geqslant \bigwedge v(\Sigma)$ as required. $\qquad \square$

The Completeness Theorem for propositional logic takes the form '$\Sigma \vDash \phi$ implies $\Sigma \vdash \phi$' and says that any statement ϕ that follows from Σ in the sense of boolean algebras, i.e. any ϕ whose valuation is at least as true as that of Σ however these valuations are chosen, actually has a formal derivation from assumptions Σ *in the formal system.* This is hardly an obvious assertion since the rules for the formal system are limited and coming up with formal proofs can be rather difficult, even for specific cases, let alone in general. It turns out that the Completeness Theorem is a strong and useful result for mathematics. In the next chapter we shall also look at what it has to say for boolean algebras in particular.

From a more philosophical point of view, the Completeness Theorem shows that any correct deduction about propositions or boolean terms can *always* be rewritten as a formal deduction in a specific formal system. By a 'correct deduction' here, we mean one that is 'semantically correct', i.e. one that can be shown to be correct in all possible cases by considering valuations into boolean algebras. There is a considerable saving here as the Completeness Theorem shows that any deduction given by any correct method can be replaced by one of a specific form using one particular proof system. In other words, the formal system is *complete* in the sense that it can make *all* necessary logical deductions about the statements that can be expressed in it. As a bonus, the rules of our formal system were chosen specifically to mimic real mathematical practice and we know them to be correct, because of the Soundness Theorem. Furthermore, these rules can be checked mechanically too.

We are now going to prove the Completeness Theorem. First, we need a lemma. Recall that a set $\Sigma \subseteq \mathrm{BT}(X)$ is *consistent* if $\Sigma \not\vdash \bot$.

Lemma 7.11 *Suppose X is a set of propositional letters, $\Sigma \subseteq \mathrm{BT}(X)$ is consistent, and $\phi \in \mathrm{BT}(X)$. Then either $\Sigma \cup \{\phi\}$ is consistent or $\Sigma \cup \{\neg \phi\}$ is consistent.*

Proof If $\Sigma \cup \{\phi\} \vdash \bot$ and $\Sigma \cup \{\neg \phi\} \vdash \bot$ then by the first of these and Reductio Ad Absurdum $\Sigma \vdash \neg \phi$. By appending a proof of $\bot$ from $\Sigma \cup \{\neg \phi\}$ (given by the second of our assumptions) on to the end of a proof for $\Sigma \vdash \neg \phi$ we obtain a proof of $\bot$ from Σ, hence Σ is inconsistent. $\square$

Theorem 7.12 (Completeness Theorem, first form) *Let X be a set, and suppose that $\Sigma_0 \subseteq \mathrm{BT}(X)$ is consistent. Then there is a valuation $v : \mathrm{BT}(X) \to \{\top, \bot\}$ such that $v(\sigma) = \top$ for all $\sigma \in \Sigma_0$.*

Proof Somehow we have to define an appropriate valuation, and this may involve making choices for the value of the letters in X. This already suggests that if X is not finite then we will need to use Zorn's Lemma.

It is not obvious how to use Zorn's Lemma directly, as decisions on the value $v(x)$ to give to a propositional letter $x \in X$ can have rather subtle consequences. So instead we use our system of formal derivations to 'control' what is going on.

We consider the set $Z = \{\Sigma \subseteq \mathrm{BT}(X) : \Sigma \supseteq \Sigma_0 \text{ and } \Sigma \nvdash \bot\}$ of all consistent extensions of Σ_0. This is a poset, where the order relation is $\subseteq$. We claim that Z has the Zorn property and so has a maximal element.

To see this, let $C \subseteq Z$ be a chain. So each element of C is a set Σ such that $\Sigma \supseteq \Sigma_0$ and $\Sigma \nvdash \bot$. Now consider $\Sigma_C = \bigcup C$. Clearly Σ_C is an upper bound for C, we just need to show that it is in Z. First, Σ_C extends Σ_0 as each element of C does. Now suppose $\Sigma_C \vdash \bot$. Then there is a single derivation p from Σ_C of $\bot$. This derivation is finite so only requires finitely many assumptions from Σ_C. Each of these assumptions comes from some $\Sigma \in C$, and as there are finitely many assumptions, only finitely many $\Sigma \in C$ are required, and of these there is a $\subseteq$-largest one, Π say, since C is a chain. Therefore the derivation p can be regarded as a derivation $\Pi \vdash \bot$, but this is impossible as $\Pi \in Z$ so by definition is consistent. Hence Σ_C is consistent.

By the previous paragraph and Zorn's Lemma there is a maximal element $\Sigma_{\max}$ in Z. Note that by Lemma 7.11 this has the property that for any $\phi \in \mathrm{BT}(X)$ either $\phi \in \Sigma_{\max}$ or $\neg\phi \in \Sigma_{\max}$, for otherwise from the lemma we could deduce that one of $\Sigma_{\max} \cup \{\phi\}$ or $\Sigma_{\max} \cup \{\neg\phi\}$ is a proper consistent extension of $\Sigma_{\max}$, contradicting maximality. Furthermore we cannot have both of ϕ, $\neg\phi$ in $\Sigma_{\max}$ since that would mean that $\Sigma_{\max}$ would be inconsistent by the Contradiction Rule. Note also that by the same argument $\top \in \Sigma_{\max}$ and $\neg\bot \in \Sigma_{\max}$ since both $\{\neg\top\}$ and $\{\bot\}$ are inconsistent.

We may now define our valuation by $v(x) = \top$ if $x \in \Sigma_{\max}$ and $v(x) = \bot$ if $\neg x \in \Sigma_{\max}$. By the previous paragraph this does indeed define a valuation and we just need to show that $v(\sigma) = \top$ for each $\sigma \in \Sigma_0$. We shall in fact show $v(\sigma) = \top$ for each $\sigma \in \Sigma_{\max}$, which suffices as $\Sigma_0 \subseteq \Sigma_{\max}$.

The proof that $v(\sigma) = \top$ for each $\sigma \in \Sigma_{\max}$ is by induction on the number of symbols in $\sigma \in \mathrm{BT}(X)$. Our induction hypothesis $H(n)$ is the statement that if σ contains at most n symbols then

$$v(\sigma) = \top \text{ if and only if } \sigma \in \Sigma_{\max}.$$

This statement is true for $n = 1$ since the only elements of $\sigma \in \mathrm{BT}(X)$ with one symbol are $\top$, $\bot$ or elements of X and these have been dealt with already by the definition of v.

If σ has $n+1$ symbols and is $\neg\tau$ for some τ then we have by the induction hypothesis $v(\tau) = \top$ if and only if $\tau \in \Sigma_{\max}$. If $\neg\tau \in \Sigma_{\max}$ then $\tau \notin \Sigma_{\max}$ since otherwise $\Sigma_{\max}$ would be inconsistent and $v(\tau) = \bot$ so $v(\neg\tau) = v(\tau)' = \top$. On the other hand if $\neg\tau \notin \Sigma_{max}$ then $\tau \in \Sigma_{max}$ by maximality of Σ_{max} and $v(\tau) = \top$ so $v(\neg\tau) = v(\tau)' = \bot$. This completes the induction step for $\neg\tau$.

If σ has $n+1$ symbols and is $(\alpha \wedge \beta)$ then by the induction hypothesis $v(\alpha) = \top$ if and only if $\alpha \in \Sigma_{max}$ and $v(\beta) = \top$ if and only if $\beta \in \Sigma_{max}$. Suppose that $(\alpha \wedge \beta) \in \Sigma_{max}$, then $\alpha, \beta \in \Sigma_{max}$ by the $\wedge$-Elimination Rule so $v(\alpha) = v(\beta) = \top$ and hence $v(\alpha \wedge \beta) = v(\alpha) \wedge v(\beta) = \top$. Alternatively, if $(\alpha \wedge \beta) \notin \Sigma_{max}$ then $\neg(\alpha \wedge \beta) \in \Sigma_{max}$ by maximality. This means that it is not the case that both $\alpha, \beta \in \Sigma_{max}$ for otherwise the following derivation would show $\Sigma_{max} \vdash \bot$ and hence one of $v(\alpha), v(\beta)$ is $\bot$ so $v(\alpha \wedge \beta) = \bot$.

Formal proof

$\neg(\alpha \wedge \beta)$	(1)	Given
α	(2)	Given
β	(3)	Given
$(\alpha \wedge \beta)$	(4)	$\wedge$-Introduction
$\bot$	(5)	Contradiction

If σ has $n+1$ symbols and is $(\alpha \vee \beta)$ then by the induction hypothesis $v(\alpha) = \top$ if and only if $\alpha \in \Sigma_{max}$ and $v(\beta) = \top$ if and only if $\beta \in \Sigma_{max}$. Suppose that $(\alpha \vee \beta) \in \Sigma_{max}$, then at least one of α, β must be in Σ_{max} for otherwise $\neg\alpha, \neg\beta \in \Sigma_{max}$ and the following would show $\Sigma_{max} \vdash \bot$.

Formal proof

$(\alpha \vee \beta)$	(1)	Given
$\neg\alpha$	(2)	Given
$\neg\beta$	(3)	Given
β	(4)	$\vee$-Elimination
$\bot$	(5)	Contradiction

So $(\alpha \vee \beta) \in \Sigma_{max}$ implies $\alpha \in \Sigma_{max}$ or $\beta \in \Sigma_{max}$ so by our induction hypothesis $v(\alpha) = \top$ or $v(\beta) = \top$ and hence $v(\alpha \vee \beta) = \top$. Alternatively, if $(\alpha \vee \beta) \notin \Sigma_{max}$, then $\neg(\alpha \vee \beta) \in \Sigma_{max}$ by maximality and we will show that this means that both of $\neg\alpha \in \Sigma_{max}$ and $\neg\beta \in \Sigma_{max}$. If not by maximality again at least one of $\alpha \in \Sigma_{max}$ or $\beta \in \Sigma_{max}$. Assume the first (the other case is similar). Then the following shows $\Sigma_{max} \vdash \bot$.

Formal proof

$\neg(\alpha \vee \beta)$	(1)	Given
α	(2)	Given
$(\alpha \vee \beta)$	(3)	$\vee$-Introduction
$\bot$	(4)	Contradiction

Therefore $v(\alpha) = v(\beta) = \bot$ so $v(\alpha \vee \beta) = \bot$ as required.

This completes the induction proof and therefore v is a valuation making all of $\Sigma_{\max}$ (and hence Σ_0) true, as required. $\qquad\square$

Theorem 7.13 (Completeness Theorem, second form) *Let X be a set, and suppose that $\Sigma \subseteq BT(X)$ and $\tau \in BT(X)$ with $\Sigma \vDash \tau$, or just $\Sigma \vDash_2 \tau$. Then $\Sigma \vdash \tau$, i.e. there is a formal derivation of τ from Σ.*

Proof Assume $\Sigma \nvdash \tau$. Then $\Sigma \cup \{\neg \tau\} \nvdash \bot$ since if otherwise we would have $\Sigma \vdash \neg\neg\tau$ by Reductio Ad Absurdum and hence $\Sigma \vdash \tau$ by $\neg$-Elimination. But then the first form of the Completeness Theorem gives us a valuation $v \colon BT(X) \to \{\top, \bot\}$ with $v(\sigma) = \top$ for all $\sigma \in \Sigma$ and $v(\neg \tau) = v(\tau)' = \top$ hence $v(\tau) = \bot$. So $\bigwedge v(\Sigma) = \top > v(\tau)$ and hence $\Sigma \nvDash \tau$. $\qquad\square$

Many of the more interesting consequences of completeness and soundness are connected with the finite nature of proofs. We can express this connection in the following important result.

Theorem 7.14 (Compactness Theorem) *Let X be an infinite set and let $\Sigma \subseteq BT(X)$. Suppose that for every finite subset $\Sigma_0 \subseteq \Sigma$ there is a valuation $v_0 \colon BT(X) \to \{\top, \bot\}$ such that $v_0(\sigma) = \top$ for all $\sigma \in \Sigma_0$. Then there is a single valuation $v \colon BT(X) \to \{\top, \bot\}$ such that $v(\sigma) = \top$ for all $\sigma \in \Sigma$.*

Proof By hypothesis, and the Soundness Theorem, every finite subset of Σ is consistent. It follows that the whole of Σ is consistent since if $\Sigma \vdash \bot$ then there is a finite proof of $\bot$ from Σ, which necessarily uses only finitely many assumptions from Σ. This would imply that some finite subset of Σ is inconsistent, contrary to hypothesis. So $\Sigma \nvdash \bot$ and by the Completeness Theorem there is a valuation as indicated. $\qquad\square$

Finally, to sum up this section, we have looked at boolean algebras in general, and the boolean algebra $2 = \{\top, \bot\}$ in particular, and proved a version of the Soundness Theorem for general boolean algebras and a version of the Completeness Theorem for the particular boolean algebra $\{\top, \bot\}$. The results we proved show that, for any set of boolean terms Σ and any other statement σ

$$\Sigma \vDash_2 \sigma \text{ implies } \Sigma \vdash \sigma$$

(the Completeness Theorem), and

$$\Sigma \vdash \sigma \text{ implies } \Sigma \vDash \sigma$$

(the Soundness Theorem). It is a triviality from the definition that

$$\Sigma \vDash \sigma \text{ implies } \Sigma \vDash_2 \sigma$$

since $2 = \{\top, \bot\}$ is an example boolean algebra that can be used in the definition of $\Sigma \models \sigma$. Thus we have proved the following.

Corollary 7.15 *For any set of boolean terms Σ and any other statement σ,*

$$\Sigma \models \sigma \text{ if and only if } \Sigma \models_2 \sigma.$$

This is hardly obvious, but the equivalence of all these statements is of great interest as it allows us to decide statements of the form $\Sigma \vdash \sigma$ by looking at valuations to the boolean algebra 2, and this is a great help as this algebra is particularly easy to compute in. In particular, from Example 7.8 we can now deduce that $p, \neg(p \wedge r) \vdash (p \vee q) \wedge r$ without having to construct a formal proof directly, and from Example 7.9 we can see that $\neg(q \wedge r), (p \vee r) \nvdash p \wedge (\neg q \vee r)$. This last fact would have been difficult to deduce directly from the formal system alone without the help of the Soundness Theorem.

7.2 Examples and exercises

We have used boolean algebras to represent the more traditional material on 'propositional logic'. The Soundness and the Completeness Theorems give a connection between the two sides of the story: *semantics* or meanings, given by boolean algebras in general or by the particular boolean algebra $2 = \{\top, \bot\}$ on the one hand, and the propositions or *boolean terms* themselves considered as strings of symbols, and *formal proofs* involving them on the other hand. The side involving boolean terms and formal proofs is useful since a proof is mechanically checkable and therefore incontrovertibly a proof of what it claims to be. The semantics side is often easier to use when it comes to proving that something is *not* derivable.

Exercise 7.16 Show that the following hold:

(i) $p \wedge (\neg p \vee q) \models q$

(ii) $\models \neg(p \wedge (q \vee r)) \vee ((p \wedge q) \vee (p \wedge r))$

(iii) $\neg p \vee q, \neg r \vee s \models \neg(p \vee r) \vee (q \vee s)$

(iv) $(p \wedge q) \wedge r \models q \wedge (r \wedge p)$

(v) $\models (\neg((\neg p \vee q) \wedge p) \vee q)$

Exercise 7.17 Determine whether

(i) $\neg(\neg q \vee r) \vdash (p \wedge r) \vee (q \wedge \neg p) \vee (\neg r \wedge q \wedge p)$

(ii) $\neg(\neg q \vee \neg p) \vdash (p \wedge q \wedge r) \vee (q \wedge \neg r \wedge \neg p) \vee (r \wedge q)$

Exercise 7.18 Our 'or' or $\vee$ should be interpreted as 'inclusive or', i.e. $a \vee b$ is 'a or b or both'. Introduce a new symbol, $+$ for *exclusive or*, where $a + b$ means 'a or b but not both at the same time'. Add the axiom

$$a + b = (a \wedge b') \vee (a' \wedge b)$$

to those for boolean algebras, and add the following formal rules of deduction to the proof system for propositional logic: 'from $(a + b)$ and $\neg a$ deduce b'; 'from $(a + b)$ and a deduce $\neg b$'; 'from $(a + b)$ and $\neg b$ deduce a'; 'from $(a + b)$ and b deduce $\neg a$'; 'from a and $\neg b$ deduce $(a + b)$'; 'from $\neg a$ and b deduce $(a + b)$'. State and prove Completeness and Soundness Theorems for the resulting system.

Why do you think I suggest the symbol '$+$' for this operation?

Definition 7.19 Two propositional terms ϕ, ψ involving propositional letters from a set X are said to be *logically equivalent* if whenever $v : X \to \{\top, \bot\}$ is a valuation then $v(\phi) = v(\psi)$. For example, the terms $(p \vee p)$ and $(p \wedge p)$ are different terms but are logically equivalent.

Exercise 7.20 Suppose that ϕ, ψ are logically equivalent. Show that $\phi \vdash \psi$, $\psi \vdash \phi$, $\phi \vDash \psi$, and $\psi \vDash \phi$.

Consider a truth table for a formula ϕ. For example, if ϕ has three proposition letters a, b, c the truth table will take the form

a	b	c	ϕ
$\top$	$\top$	$\top$	v_1
$\top$	$\top$	$\bot$	v_2
$\ldots$	$\ldots$	$\ldots$	$\ldots$
$\bot$	$\bot$	$\bot$	v_8

Exercise 7.21 By considering the rows of the table for which $v_i = \top$ show that ϕ is logically equivalent to an 'or' or *disjunction*

$$\tau_1 \vee \tau_1 \vee \ldots \vee \tau_k$$

where each τ_j is an 'and' or *conjunction* $\tau_{j,1} \wedge \tau_{j,2} \wedge \tau_{j,3}$ and $\tau_{j,1}$ is a or $\neg a$, $\tau_{j,2}$ is b or $\neg b$, and $\tau_{j,3}$ is b or $\neg c$.

More generally, show that any propositional formula ϕ is logically equivalent to one, ψ, in *Disjunctive Normal Form (DNF)*, meaning ψ is a disjunction

$$\tau_1 \vee \tau_1 \vee \ldots \vee \tau_k$$

where each τ_j is a conjunction $\tau_{j,1} \wedge \tau_{j,2} \wedge \ldots \wedge \tau_{j,l}$ and each $\tau_{j,i}$ is a propositional letter or the negation of a propositional letter.

Exercise 7.22 By considering the rows of the truth table for which $v_i = \bot$, or otherwise, show that ϕ is also logically equivalent to a formula ψ in *Conjunctive Normal Form (CNF)*

$$\tau_1 \wedge \tau_1 \wedge \ldots \wedge \tau_k$$

where each τ_j is a disjunction $\tau_{j,1} \vee \tau_{j,2} \vee \ldots \vee \tau_{j,l}$ and each $\tau_{j,i}$ is a propositional letter or the negation of a propositional letter.

Exercise 7.23 Suppose Σ is a set of terms in $\mathrm{BT}(X)$ and there is a boolean algebra B such that for each finite subset $\Sigma_0 \subseteq \Sigma$ there is a valuation $v \colon X \to B$ with $\bot < \bigwedge v(\Sigma_0)$. Show that there is another valuation $w \colon X \to \{\bot, \top\}$ such that $w(\sigma) = \top$ for each $\sigma \in \Sigma$. (Hint: apply the Soundness Theorem to show Σ is consistent, and then use completeness.)

The next exercise presents a notion of 'positive' boolean terms. (Compare with Definition 4.12.)

Exercise 7.24 Add the operation $\to$ for 'implies' to the set of logical symbols $\wedge, \vee, \neg, \top, \bot$ for propositional logic. A term involving these symbols together with propositional letters from a set X is said to be *positive* if neither of the symbols $\bot$ or $\neg$ appear in it. Prove (by induction on terms) that if ϕ is a positive term and v is the valuation sending each propositional letter x to $\top$, then $v(\phi) = \top$. Deduce that $\neg p$ is not logically equivalent to a positive term.

Exercise 7.25 Prove a converse to the previous exercise by showing that if a propositional term ϕ has the property that the valuation v sending each propositional letter x to $\top$ gives $v(\phi) = \top$, then ϕ is logically equivalent to a positive term. (Hint: use induction on the number of distinct propositional letters in ϕ. In particular show that $\phi(p_1, p_2, \ldots, p_n)$ is logically equivalent to a statement of the form

$$(p_{j_1} \vee p_{j_2} \vee \ldots \vee p_{j_k}) \wedge \bigwedge_i (p_i \to \phi(p_1, \ldots, p_{i-1}, \top, p_{i+1}, \ldots, p_n)).\,)$$

Exercise 7.26 Suppose that $\Sigma \vdash \phi$ where Σ is a set of positive propositional terms Show that ϕ is also logically equivalent to a positive term. (Hint: consider the valuation v sending each propositional letter to $\top$. What does the Soundness Theorem say about $\Sigma \vdash \phi$ and v? Now apply the characterisation of positive terms in the previous exercise.)

Finally, we shall show that with infinitely many propositional letters, our propositional logic can simulate the systems of Chapters 3 and 4.

Example 7.27 Let X be the set 2^* of strings of 0s and 1s of finite length, considered as a set of propositional letters. To avoid any possibility of confusion we shall write a string σ in 2^* as q_σ when we think of it as a propositional letter. Suppose $\Sigma \subseteq 2^*$ is a set of given strings or axioms for the system of Chapter 3. We define a set Γ_Σ of terms in $BT(X)$ that represent Σ and the proof system of Chapter 3.

The set Γ_Σ is the set of: q_σ for each $\sigma \in \Sigma$; the statement $\neg q_{\tau 0} \vee \neg q_{\tau 1} \vee q_\tau$ for each $\tau \in 2^*$; and the statement $\neg q_\rho \vee (q_{\rho 0} \wedge q_{\rho 1})$ for each $\rho \in 2^*$. The idea is that these propositional formulas represent the given statements and Shortening and Lengthening Rules of the previous system. Indeed, it is the case that $\Sigma \vdash \tau$ in the system of Chapter 3 if and only if $\Gamma_\Sigma \vdash q_\tau$ in propositional logic.

One direction is straightforward: by induction on the length of proofs in the system of Chapter 3, if $\Sigma \vdash \tau$ then $\Gamma_\Sigma \vdash q_\tau$. We leave this as an exercise.

For the other direction, suppose $\Sigma \nvdash \tau$. Then by the Completeness Theorem for that system, Theorem 3.12, there is an infinite path p in 2^* such that $\tau \in p$ but $p \cap \Sigma = \varnothing$. Define a valuation $v: X \to \{\top, \bot\}$ by $v(q_\rho) = \bot$ if $\rho \in p$ and $v(q_\rho) = \top$ if $\rho \notin p$. Then $v(q_\rho) = \top$ for all $\rho \subset \Sigma$ since $\Sigma \cap p = \varnothing$; $v(\neg q_{\tau 0} \vee \neg q_{\tau 1} \vee q_\tau) = \top$ since if $v(q_\tau) = \bot$ then τ is in the path p and so at least one of $\tau 0, \tau 1$ is in the path and hence $v(\neg q_{\tau 0}) = \top$ or $v(\neg q_{\tau 1}) = \top$; and $v(\neg q_\rho \vee (q_{\rho 0} \wedge q_{\rho 1})) = \top$ since if $v(q_{\rho 0} \wedge q_{\rho 1}) = \bot$ then one of $\rho 0$ or $\rho 1$ is in the path p hence ρ itself is in p and so $v(\neg q_\rho) = \top$. Thus v makes all statements in Γ_Σ true and $v(q_\tau) = \bot$ since $\tau \in p$, so this shows that $\Gamma_\Sigma \nvdash q_\tau$ by the Soundness Theorem for propositional logic.

Example 7.28 As a converse to the last example, *and for ambitious readers only*, we shall try to show that the system of Chapter 3 can be used as a basis for propositional logic.

Once again, we take infinitely many propositional letters p_i, with i ranging over natural numbers, but we let $X = \{p_i : i \in \mathbb{N}\}$ and look at $BT(X)$, the set of propositional statements involving the p_i. It turns out that $BT(X)$ is also countable, so the set of all terms in $BT(X)$ can be enumerated as σ_i where i ranges over $\mathbb{N}$.

We consider a boolean term θ from $BT(X)$ and we try to determine whether θ is possible or impossible. (This is rather more general than the discussion at the end of Section 3.1, since we are specifying a collection of boolean terms, and not a collection of sets of values. However, an arbitrary finite set of values, for example $p_0 = 0$, $p_1 = 1$, $p_2 = 1$, $p_3 = 0$, ..., $p_{k-1} = 0$ can be specified as a single boolean term as for example $\neg p_0 \wedge p_1 \wedge p_2 \wedge \ldots \wedge \neg p_{k-1}$.)

We also need to consider an additional set Ξ of situations that we know to be impossible, based on the logic of $\neg$, $\wedge$ and $\vee$. These are as follows.

- The sets $\{\bot\}$ and $\{\neg \top\}$ are in Ξ, as no situation in which $\bot$ or $\neg \top$ holds is possible.
- For every $\phi \in BT(X)$, the set $\{\phi, \neg \phi\}$ is in Ξ, as no situation in which both ϕ and $\neg \phi$ hold is possible.
- For every $\phi, \psi \in BT(X)$, the set $\{(\phi \vee \psi), \neg \phi, \neg \psi\}$ is impossible and is in Ξ.
- For every $\phi, \psi \in BT(X)$, both $\{\neg(\phi \vee \psi), \phi\}$ and $\{\neg(\phi \vee \psi), \psi\}$ are impossible and are in Ξ.
- For every $\phi, \psi \in BT(X)$, both $\{(\phi \wedge \psi), \neg \phi\}$ and $\{\neg(\phi \wedge \psi), \neg \psi\}$ are impossible and are in Ξ.
- For every $\phi, \psi \in BT(X)$, the set $\{\neg(\phi \wedge \psi), \phi, \psi\}$ is impossible and is in Ξ.

(These rules are sometimes known as the tableau rules for propositional logic, and a proof system derived from them is a tableau system.)

Now, we identify each of our finite sets $S \subseteq BT(X) = \{\sigma_i : i \in \mathbb{N}\}$ with a set S^* of strings $s_0 s_1 \ldots s_{k-1}$ of minimal length k corresponding to one more than the largest index of $\sigma_i \in S$, where

$$S^* = \left\{ s_0 s_1 \ldots s_{k-1} \in \{0, 1\}^k : \forall i < k (\sigma_i \in S \rightarrow s_i = 1) \right\}$$

so that, if S does not contain σ_i for some $i < k$, then both possible values of s_i are allowed. We will do this for each $S \in \Xi$ and also apply this to the singleton set $\{\neg \theta\}$.

This then gives a large set of strings $\Pi_\theta = \{\neg \theta\}^* \cup \bigcup \{S^* : S \in \Xi\}$. The main property of this is that $\Pi_\theta \vdash \bot$ in the system of Chapter 3 if and only if $\theta \vdash \bot$ in propositional logic. One direction is proved by induction on the length of derivations in the system of Chapter 3. The other direction is proved by saying that if $\Pi_\theta \nvdash \bot$ then there is a path p through the tree 2^* avoiding Π_θ, and this path defines a valuation $v : BT(X) \rightarrow \{\top, \bot\}$ with $v(\sigma_i) = \top$ if some string $s_0 s_1 \ldots s_i$ is in p with $s_i = 1$. It is not trivial in this case to see that this is a valuation, i.e. satisfies $v(\bot) = \bot$, $v(\top) = \top$, $v(\neg \phi) = \neg v(\phi)$, $v(\phi \wedge \psi) = v(\phi) \wedge v(\psi)$, and $v(\phi \vee \psi) = v(\phi) \vee v(\psi)$, because v is defined on all formulas of $BT(X)$ and not just on X, but this follows from the fact that the path p avoids all S^* with $S \in \Xi$. This valuation clearly makes $\neg \theta$ false, so makes θ true. Thus $\theta \nvdash \bot$.

Full details are left as an exercise, as is the extension of this to the characterisation of the provability $R \vdash \tau$ of a statement τ from a set of statements R in propositional logic.

Example 7.29 Let X be a non-empty set and Σ a set of statements of the form $a \prec b$ or $a \not\prec b$ for elements $a, b \in X$. We will define a translation $t(\sigma)$ of statements σ in the system of Chapter 4 and a set Ξ of statements such that for $t(\Sigma) = \{t(\sigma) : \sigma \in \Sigma\}$ we have: $\Sigma \vdash \sigma$ holds in the sense of Chapter 4 if and only if $\Xi \cup t(\Sigma) \vdash t(\sigma)$.

Let Y be the set of proposition letters $p_{a,b}$ for all $a, b \in X$, and translate statements of the system of Chapter 4 by: $t(\bot) = \bot$; $t(a \prec b) = p_{a,b}$; and $t(a \not\prec b) = \neg p_{a,b}$. Now define Ξ to be the set of statements: $\neg p_{a,b} \vee \neg p_{b,c} \vee p_{a,c}$ for all $a, b, c \in X$; and $p_{a,a}$ for each $a \in X$. We now want to prove that $\Sigma \vdash \sigma$ if and only if $\Xi \cup t(\Sigma) \vdash t(\sigma)$. One direction, that $\Sigma \vdash \sigma$ implies $\Xi \cup t(\Sigma) \vdash t(\sigma)$, is a straightforward induction on the length of derivations in the system of Chapter 4 and is left as an exercise.

For the other direction, suppose $\Sigma \nvdash \sigma$. Then by the Completeness Theorem of Chapter 4, Theorem 4.10, there is a partial order $<$ on X making Σ true and σ false. Define a valuation by $v(p_{a,b}) = \top$ if $a < b$ holds in this order, and $\bot$ otherwise. So $v(t(\tau)) = \top$ for all $\tau \in \Sigma$ and $v(t(\sigma)) = \bot$. Also $v(\rho) = 1$ for all $\rho \subset \Xi$, by the transitivity and irreflexivity axioms for a partial order. Thus v makes $\Xi \cup t(\Sigma)$ true and $t(\sigma)$ false and so $\Xi \cup t(\Sigma) \nvdash t(\sigma)$, as required.

7.3 The complexity of satisfiability*

We have seen that, given a set of propositional letters X, a finite set of propositional formulas Σ in X and a further formula σ, there are algorithms to decide whether $\Sigma \vdash \sigma$. As pointed out in Section 6.3, one algorithm goes by directly searching for a proof according to some strategy that is guaranteed to give an answer. From the results in this chapter, $\Sigma \vdash \sigma$ is equivalent to $\Sigma \vDash_2 \sigma$, so another quite different algorithm goes by checking all possible valuations $v : X \rightarrow \{\top, \bot\}$ and checking that each valuation v that makes each statement in Σ true also makes σ true. Thus we have two quite different algorithms for deciding whether $\Sigma \vdash \sigma$ and there are others too.

Just knowing that there is an algorithm to solve our problem is not the end of the story. We might also like to ensure that our algorithm runs efficiently and is effective in the sense that it runs in reasonable time for typical inputs. But here there is a problem.

If X has n propositional letters, then there are 2^n different valuations on the letters in X. Even if n is a very reasonable (and typical) 100, that would mean that there are 2^{100} individual valuations to check, and this is an unreasonably large number to expect any sort of computer to run through in a human lifetime. And of course $n = 1000$ would be much much worse.

The other algorithm does not fare any better. Suppose we consider a special case (but a rather typical one) where we ask whether $\tau \vdash \bot$, where τ is $\tau_1 \wedge \tau_2 \wedge \ldots \wedge \tau_k$ and each τ_i is a disjunction of three atomic or negated atomic statements $(\tau_{i1} \vee \tau_{i2} \vee \tau_{i3})$. Then our proof strategies ask us to decide whether $\tau_1, \tau_2, \ldots, \tau_k \vdash \bot$ and this in turn requires us to check each $\tau_{1j_1}, \tau_{2j_2}, \ldots, \tau_{kj_k} \vdash \bot$ for each possible combination of $j_m \in \{1, 2, 3\}$. There are then 3^k such proofs to check, and this is unreasonably large for any computer, even for reasonable values of k of the order of 100.

Of course these arguments prove nothing on their own about how hard it is to decide whether $\Sigma \vdash \sigma$ or whether $\tau \vdash \bot$, since these are just two possible algorithms; there may be other algorithms for the same problem and one of those may be much better than the two we have available at present. To current knowledge, this could indeed be the case, but most experts in the field seem to think it unlikely.

To study this problem more thoroughly will take us into the subject of computability, and in particular *complexity theory*, which tries to study the inherent complexity of a problem such as $\Sigma \vdash \sigma$ in terms of the performance of the best-possible algorithm to solve it. I will not give the official definitions of 'algorithm', or an algorithm's 'performance' here as it would take us too far off track, but I will say enough to give a flavour of the subject and the main open problem.

In complexity theory, a *problem* is a set of inputs, usually encoded as a string of symbols from a finite set of symbols, to a question, which should have a definite yes/no answer. A problem can also be thought of as that set of input data for which the answer to the question is yes. Thus 'does $\Sigma \vdash \sigma$?' is a typical problem, where the inputs are the data Σ, σ. This problem is usually simplified to special cases that nevertheless contain all of the difficulties. These are given in the following definition.

Definition 7.30 The problem *SAT* is the set of *satisfiable* propositional formulas, i.e. formulas σ in some set of letters X for which there is a valuation $v \colon \mathrm{BT}(X) \to \{\top, \bot\}$ with $v(\sigma) = \top$.

The problem *TAUT* is the set of *tautologies*, i.e. formulas σ in some set of letters X for which every valuation $v \colon \mathrm{BT}(X) \to \{\top, \bot\}$ has $v(\sigma) = \top$.

These problems are related. σ is in TAUT if and only if $\neg \sigma$ is not in SAT, and $\Sigma \vdash \sigma$ if and only if $(\neg \bigwedge \Sigma) \vee \sigma$ is in TAUT, which is the case if and only if $(\bigwedge \Sigma) \wedge (\neg \sigma)$ is not in SAT. The algorithms described earlier can easily be adapted to solve SAT and TAUT, with similar difficulties about the time that they take.

In complexity theory, the time taken by an algorithm is regarded as a function of the number of symbols in the input. This length of the input is denoted n and for the problems we are considering here, SAT and TAUT, n is the number of $\vee, \wedge, \neg, (,), \top, \bot$ and propositional letters in the input. It is a non-obvious but interesting fact that as far as theoretical running time is concerned it does not much matter what kind of (theoretical or actual) computer an algorithm runs on. Within a constant multiplying factor, or very occasionally a squaring or a square-root, all normal deterministic computers run the same algorithms in the same time. A very fast algorithm might have running time which is $O(n)$, a slightly slower might have running time $O(n^2)$. There are also $O(n^3)$, $O(n^4)$, ..., algorithms, and these may also be useful in practice. Generally speaking, it is conventional to say that an algorithm is useful or *effective* if its running time is the order of a polynomial in n. Of course this is not the end of the story, and one might want to know the order of the polynomial, size of the coefficients, etc., but an algorithm which does *not* have running time the order of a polynomial in n seems pretty clearly one that will not be effective, practical or useful in general.

Definition 7.31 The class P of problems soluble in *polynomial time* is the class of problems for which there is a deterministic algorithm solving it in polynomial time, i.e. having running time which is $O(n^k)$ for some constant k depending on the problem, but not the input to the problem.

To estimate the performance of our algorithms for SAT and TAUT, we need to relate the length of input n to the number of propositional letters. In the very worst case, a formula may be something like $((((a \wedge b) \vee c) \wedge d). . .)$ with around $n/4$ distinct propositional letters. Of course the SAT and TAUT problems for such formulas may be easy to solve, but rather difficult examples can easily be constructed with around $\sqrt{n}$ propositional letters. Thus (since we are only interested in running times of the order of a polynomial in n anyway, and square-roots or constant factors are immaterial) we might as well make the simplification that n is the same as the number of propositional letters. Then we see straight away that the running time of our algorithms is closer to 2^n than a polynomial in n, since for example there are 2^n many valuations to consider for a single formula. Thus our algorithms are definitely not running in polynomial time.

This does not answer our basic question, however.

Question 7.32 *Are SAT and TAUT in the class P of polynomial-time soluble problems?*

This question is a version of the famous and unsolved 'P = NP?' problem, which is arguably one of the most interesting and important (and possibly one of the most difficult) open problems in mathematics at the current time. To explain the connection we need to say what NP and co-NP are, and what they have to do with SAT and TAUT.

To give an algorithm for SAT, we can run through all valuations $v: X \rightarrow$ $\{\top, \bot\}$ and see whether any of them make our formula true. Given a valuation, it is easy and fast to check it against a formula. But if the valuation happens to make the formula true, we do not need to check any more valuations, we can just answer 'yes'. (Of course, if the formula is not in SAT we still need to check all the valuations.) Now suppose our machine was given the freedom to make guesses all on its own, and also somehow given an ability always to choose the lucky option when one is available. Then SAT would be computable in polynomial time: our machine is set off on input σ, and (assuming σ is in SAT) very luckily chooses the right valuation straight away, verifies it correct, and answers 'yes'. On the other hand (assuming σ is not in SAT) our machine tries its best to be lucky in choosing a valuation, but when verifying it discovers this valuation is incorrect, and so must answer 'no'.

Such machines are called *non-deterministic*. It is important to realise that verifying the lucky guess is an essential part of a non-deterministic machine, since *to the machine*, it is the 'yes' answer that matters, and not our intentions when programming the machine. It is like a mouse running through a maze looking for a lump of cheese. The mouse only wants to take the correct turns to reach the cheese, and does not understand who built the maze or what the maze is for; in the same way the non-deterministic machine's luck leads it to a 'yes' if its program allows it to do so. This is why we still need to program in a verification step to ensure that it only reaches 'yes' when we want it to.

Definition 7.33 The class NP is the class of problems soluble in polynomial time on a non-deterministic computer.

We have seen that SAT is in NP. The 'P = NP?' question asks whether *every* NP problem is soluble in polynomial time on a deterministic machine.

Question 7.34 *Does* P = NP?

This sounds as if it would be much harder to prove that P = NP instead of just showing SAT is in P, but in fact a result due to Stephen Cook states that SAT $\in$ P implies that P $=$ NP; in other words, SAT is in some sense the hardest member of NP (or rather, equal-hardest with some other problems in NP). Cook's Theorem is proved by describing non-deterministic computers in the propositional logic, not so very far removed from the methods used in Examples 7.27 and 7.29, and because SAT therefore encodes the computations of non-deterministic polynomial-time computers, the terminology used is that SAT is *NP-complete*. Hundreds of other NP-complete problems have been discovered, including ones of great importance in optimisation, number theory and cryptography, so knowing whether they are in P or not is of key importance. Unfortunately, this problem seems too difficult at the present.

There is a subtle asymmetry between 'yes' and 'no' in a non-deterministic computer. If we change these round, we get a different model of a computer, and the class of co-NP problems.

A *co-nondeterministic* computer is like a non-deterministic one, except this time it is the 'no' answer that is regarded as lucky, and the computer will make guesses that leads to a 'no' wherever possible. The problem TAUT is easy to solve on a co-nondeterministic computer. Once again, we guess a valuation, but this time hoping for a valuation that will show the input is not a tautology. If we are lucky, our valuation will make the formula false and our verification will show this to be the case. In this case we can answer 'no'. If we are not lucky we have no choice but to answer 'yes'.

Definition 7.35 The class co-NP is the class of problems soluble in polynomial time on a co-nondeterministic computer.

It is fairly easy to see that P $-$ NP if and only if P $=$ co-NP, and that (by Cook's Theorem) these are equivalent to the statement 'TAUT $\in$ P'. Two further interesting problems are,

Question 7.36 *Does* NP $=$ co-NP?

Question 7.37 *Does* NP $\cap$ co-NP $=$ P?

It is not difficult to show that if P $=$ NP then NP $=$ co-NP and NP $\cap$ co-NP $=$ P, but no converses to these implications are expected. Thus the two questions above are expected to have negative answers but appear to be harder to solve than 'P $=$ NP?' itself. In fact about the only easy thing in this subject seems to be to invent unsolved problems! It is not expected that TAUT $\in$ NP or SAT $\in$ co-NP, though these are open questions equivalent to NP $=$ co-NP. The class NP $\cap$ co-NP contains interesting problems such as problems to do with factorisation that are not known to be in P.

8

Filters and ideals

8.1 Algebraic theory of boolean algebras

In this chapter we start to explore the theory of boolean algebras as an algebraic theory in its own right, in a way analogous to ring theory, say. We will see many applications of the Completeness and Soundness Theorems proved in the last chapter.

We start with an important definition concerning boolean algebras.

Definition 8.1 Let B, C be boolean algebras. A *homomorphism* from B to C is a map $h: B \to C$ such that, for all $a, b \in B$,

- $h(a \vee b) = h(a) \vee h(b)$
- $h(a \wedge b) = h(a) \wedge h(b)$
- $h(\top) = \top$
- $h(\bot) = \bot$
- $h(a') = h(a)'$

Here, $\vee$ and $\wedge$, etc., are calculated inside B on the left hand side, and inside C on the right. In fact, the last condition (on complementation) is not necessary and follows from the other four, since if those four hold then we have $\top = h(\top) = h(a' \vee a) = h(a') \vee h(a)$ and $\bot = h(\bot) = h(a' \wedge a) = h(a') \wedge h(a)$ so $h(a') = h(a)'$ by Proposition 5.22 on the Uniqueness of Complements.

We will see several examples of homomorphisms later, but first we study one particular homomorphism of boolean algebras that applies to all such B.

Definition 8.2 Let B be a boolean algebra. Define a boolean algebra $B^{\mathrm{op}} = (B, \wedge^{\mathrm{op}}, \vee^{\mathrm{op}}, ', \top^{\mathrm{op}}, \bot^{\mathrm{op}})$ with the same underlying set and the same complementation operation as B, by defining the other operations as follows

- $x \wedge^{\mathrm{op}} y = x \vee y$

100

- $x \vee^{op} y = x \wedge y$
- $\top^{op} = \bot$
- $\bot^{op} = \top$

B^{op} is a boolean algebra in its own right, called the *opposite* of B. The map $x \mapsto x'$ is a homomorphism $B \to B^{op}$ which is also one-to-one and onto. In other words it is an *isomorphism* of boolean algebras. This 'opposite' map turns boolean algebras 'upside down' swapping $\top$ and $\bot$ and reversing the order relation. It also swaps $\wedge$ and $\vee$, showing that we should expect these operations to have similar properties. More importantly, it shows that for every notion in a boolean algebra there is a *dual* notion, in which the order is reversed and $\top$, $\bot$ and $\wedge$, $\vee$ are swapped over, and proofs of propositions for one kind carry over directly to the dual. Again, we will see plenty of examples in this chapter.

As well as being used to model propositional logic, boolean algebras are closely related to rings. (See the exercises below for more details.) In ring theory, the important subsets of a ring are the ideals. These are the subrings which can be used to 'quotient out' the ring to get a new (and often nicer) ring. One way of understanding ideals of rings is that they are the kernels of homomorphisms. The same applies for boolean algebras.

Definition 8.3 Let $h: B \to C$ be a homomorphism of boolean algebras. Then the *kernel* of h is the subset $\ker h = \{a \in B : h(a) = \bot\}$.

Proposition 8.4 *Let $h: B \to C$ be a homomorphism of boolean algebras and $I = \ker h$. Then I satisfies: (a) if $a \in I$ and $b \in I$ then $a \vee b \in I$; (b) if $a \in I$ and $b \in B$ with $b \leqslant a$ then $b \in I$. In particular, it follows from (b) that I contains $\bot$ and is closed under $\wedge$.*

Proof If $a \in I$ and $b \in I$ then $h(a) = h(b) = \bot$ so $h(a \vee b) = h(a) \vee h(b) = \bot$, so $a \vee b \in I$. Also, if $a \in I$ and $b \in B$ with $b \leqslant a$ then $h(b) = h(b \wedge a) = h(b) \wedge h(a) \leqslant \bot$ so $h(b) = \bot$ and $b \in I$. $\qquad\square$

Definition 8.5 Let B be a boolean algebra, or more generally, a lattice. An *ideal* of B is a non-empty subset $I \subseteq B$ such that for all $x, y \in B$: (a) $x \leqslant y \in I$ implies $x \in I$; and (b) $x, y \in I$ implies $x \vee y \in I$. The ideal I is *proper* if it is not equal to the whole of B.

As you may have guessed, there is a similar dual notion obtained by replacing $\wedge$, $\vee$, $\top$, $\bot$ by their duals $\vee$, $\wedge$, $\bot$, $\top$. The dual notion to that of an ideal is called a filter.

Definition 8.6 A *filter* of B is a non-empty subset $F \subseteq B$ such that for all $x, y \in B$: (a) $x \geqslant y \in F$ implies $x \in F$; and (b) $x, y \in F$ implies $x \wedge y \in F$. It is *proper* if it is not equal to the whole of B.

As mentioned, filters and ideals are dual concepts. If I is an ideal then $F = \{a' : a \in I\}$ is the corresponding filter, and *vice versa*.

The way to picture these is to think of an ideal as representing a set of elements to be thought of as 'negligible', 'small' or 'false'. If $x \leqslant y$ and y is 'false' then x should also be false as it is 'more false' than y. Also if $x, y \in I$ are both 'false' then so is $x \vee y$ or 'x or y'. Similarly a filter is a set of elements that can be thought of as 'large' or 'true'. If $x \geqslant y$, i.e. x is 'more true' than a true statement y, then x should be 'true', and if both x, y are 'true' then so should 'x and y' or $x \wedge y$. In this picture, we should only be interested in proper ideals and proper filters, because we want to retain at least *some* distinction between 'true' and 'false'.

Another way of saying the same thing – a way that is algebraically somewhat more sophisticated – is to say that we can factor a boolean algebra out by an ideal to make everything in the ideal look like $\bot$. Similarly we can factor out by a filter to make everything in the filter look like $\top$.

Factoring out an ideal or filter has repercussions on the whole of the boolean algebra, though. If $x, y \in B$ and I is an ideal consisting of objects to be thought of as 'false' or 'negligible' then x and y should be regarded as *equivalent modulo the ideal I* if the terms representing how x and y differ, i.e. the terms $x \wedge y'$ and $x' \wedge y$, are both negligible. (In the case when B is a boolean algebra of subsets of some set X you can think of this as saying the set differences $x \setminus y$ and $y \setminus x$ are both negligible.)

Exercise 8.7 (For those who have read Section 5.3.) Switching when necessary between B as a boolean algebra and B as a boolean ring, the difference between x and y is $x - y$, or (as B has characteristic 2), $x + y$. But $x + y = (x \wedge y') \vee (x' \wedge y)$, and so $x - y \in I$ if and only if $(x \wedge y') \vee (x' \wedge y) \in I$, which is the case if and only if both $(x \wedge y'), (x' \wedge y) \in I$, since ideals are closed under $\vee$.

Dually, x and y are *equivalent modulo the filter F* if $x \wedge y'$ and $x' \wedge y$ are in the corresponding ideal, i.e. $(x \wedge y')' \in F$ and $(x' \wedge y)' \in F$, which is equivalent to saying that both $x' \vee y \in F$ and $x \vee y' \in F$. (You can think of this as saying that the filter makes both implications $x \to y$ and $y \to x$ true.)

Definition 8.8 Let B be a boolean algebra and $I \subseteq B$ an ideal. Then B/I is the set of equivalence classes for the equivalence relation

$$a \sim b \text{ if and only if } (a' \wedge b) \in I \text{ and } (a \wedge b') \in I.$$

Similarly, if $F \subseteq B$ a filter. Then B/F is the set of equivalence classes for the equivalence relation

$$a \sim b \text{ if and only if } (a \vee b') \in F \text{ and } (a' \vee b) \in F.$$

The quotients B/I and B/F are made into a boolean algebra by defining

- $\top = [\top] = F$
- $\bot = [\bot] = I$
- $[a] \vee [b] = [a \vee b]$
- $[a] \wedge [b] = [a \wedge b]$
- $[a]' = [a']$

for each $a, b \in B$, where $[x]$ denotes the $\sim$-equivalence class of x.

Proposition 8.9 *In each of these cases, $\sim$ is an equivalence relation, the operations of $\vee$, $\wedge$ and $'$ on the quotient are well defined (i.e. do not depend on the choice of the representative of the equivalence class used to define them) and satisfy the axioms for a boolean algebra. The quotient algebra is proper (i.e. not degenerate) if and only if the ideal (or filter) used is proper.*

Proof This is a long detailed check against the axioms. We will do some typical steps, including the more difficult ones, and leave the remaining parts as an exercise for the more energetic students.

To see that $\sim$ is an equivalence, note that $x \sim x$ as $\bot = x \wedge x' = x' \wedge x \subseteq I$. The symmetry axiom is obvious. For transitivity, suppose $x \wedge y'$, $x' \wedge y$, $y \wedge z'$, $y' \wedge z$ are all in I. Then

$$x \wedge z' = (x \wedge z' \wedge y) \vee (x \wedge z' \wedge y') = (x \wedge z') \wedge ((y \wedge z') \vee (x \wedge y'))$$

which is in I as $(y \wedge z')$, $(x \wedge y') \in I$ and I is closed downwards and under $\vee$. Similarly for $x' \wedge z$.

The operations on B/I are well defined. For example if $x_1 \sim x_2$ and $y_1 \sim y_2$ then $x_1 \wedge x_2'$, $x_1' \wedge x_2$ and $y_1 \wedge y_2'$, $y_1' \wedge y_2$ are all in I. Thus $(x_1 \wedge y_1) \wedge (x_2 \wedge y_2)' = x_1 \wedge y_1 \wedge (x_2' \vee y_2') = (x_1 \wedge y_1 \wedge x_2') \vee (x_1 \wedge y_1 \wedge y_2') = (x_1 \wedge y_1) \wedge ((x_1 \wedge x_2') \vee (y_1 \wedge y_2'))$ which is in I by the closure properties of I. All other cases are similar.

Checking the axioms for a boolean algebra is now easy. (I suggest the easiest way is to check against the properties in Proposition 5.15; see also Proposition 5.16.) To check that the quotient is proper, it suffices to check that it is not the case that $\top \sim \bot$. But if $\top \sim \bot$ then $\top = \top \wedge \bot' \in I$, so $I = B$ is not proper. $\square$

In the sequel, I will tend to concentrate on filters rather than ideals since (being an optimist) I prefer to focus on true statements rather than false ones, but as we have seen these ideas are interchangeable. (Is this boolean algebra half-true or half-false?) Pessimists can easily translate what I have to say to ideals using duality.

The next lemma presents a useful way of making a new filter from an old one by adding a new element x to the filter and closing the resulting set under $\leqslant$ and $\wedge$.

Lemma 8.10 *Let B be a boolean algebra, $x \in B$, and $F \subseteq B$ a filter of B. Then*

$$G = \{g : g \geqslant x \wedge f, \text{ for some } f \in F\}$$

is a filter containing both x and all elements of F, and is in fact the least such filter. The filter G is a proper filter except when $x' \in F$.

Proof Checking that G is a filter is routine. If $g \in G$ and $h \geqslant g$ then it is clear that $h \in G$ from the definition. If $g_1 \geqslant x \wedge f_1$ and $g_2 \geqslant x \wedge f_2$ with $f_1, f_2 \in F$ then $g_1 \wedge g_2 \geqslant (x \wedge f_1) \wedge (x \wedge f_2) = x \wedge (f_1 \wedge f_2)$ and $f_1 \wedge f_2 \in F$ since F is a filter.

For the final part, suppose that G is not proper. Then $\bot \in G$ so $\bot \geqslant x \wedge f$ for some $f \in F$ so $f \leqslant x'$. Hence $x' \in F$ as F is a filter. $\qquad\square$

Our set $\mathrm{BT}(X)$ of boolean terms over X is close to being a boolean algebra in its own right. It has operations of $\wedge$, $\vee$ and complementation given by $t, s \mapsto (t \wedge s)$, $t, s \mapsto (t \vee s)$, $t \mapsto \neg t$, for terms $t, s \in \mathrm{BT}(X)$. The problem is that certain distinct terms should be 'equal'. For example, for x, $(x \wedge x)$ and $(x \vee x)$ ought to be equal to each other as elements of the boolean algebra but are in fact distinct terms. The solution is once again to use an equivalence relation.

Lemma 8.11 *Let X be a set. Then we may define an equivalence relation $\sim$ on $\mathrm{BT}(X)$ by*

$$t \sim s \text{ if and only if } t \vdash s \text{ and } s \vdash t.$$

The set of equivalence classes $[t]$ of $\mathrm{BT}(X)$ forms a boolean algebra when we define

- $[t] \leqslant [s]$ *if and only if $t \vdash s$*
- $[t] \wedge [s] = [(t \wedge s)]$
- $[t] \vee [s] = [(t \vee s)]$
- $[t]' = [\neg t]$
- $\top = [\top]$

- $\bot = [\bot]$

Proof This is more axiom-checking for energetic students! That $\sim$ is an equivalence is an easy consequence of the proof rules for propositional logic. Note that the operations are also well defined, for if $t_1 \sim t_2$ and $s_1 \sim s_2$ then $t_1 \vdash t_2$, $s_1 \vdash s_2$ and so, by $\wedge$-Elimination and $\wedge$-Introduction, $t_1 \wedge s_1 \vdash t_2 \wedge s_2$. The other direction, all the other cases, and the boolean algebra axioms are all proved in the same way by construction of propositional logic proofs. $\qquad\square$

Definition 8.12 The boolean algebra defined in the previous lemma over the set X is called the *free boolean algebra* over X and is denoted free(X).

The free boolean algebra contains as few identifications of boolean terms as possible – that is why it is called 'free'. Only terms that *must* be identified because of proofs in the propositional calculus are actually identified. In fact any boolean algebra is a quotient of a free algebra by a filter – see Exercise 8.23.

In the next result, we use filters in the free boolean algebra to measure 'at least as true as'.

Theorem 8.13 (Another version of the Soundness Theorem) *Let X be a set and consider the valuation $v\colon X \to$ free(X) defined by $v(x) = x$. Then for $\Sigma \subseteq BT(X)$ and $\sigma \in BT(X)$ we have: if $\Sigma \vdash \sigma$ then every filter G of free(X) that contains $v(\Sigma)$ also contains $v(\sigma)$.*

Proof Let G be a filter of free(X). Define a homomorphism $w\colon$ free$(X) \to$ free$(X)/G$ by $x \mapsto x/G$, where x/G is the equivalence class of x in the quotient algebra, and let $u\colon BT(X) \to$ free$(X)/G$ be the composition of v and w. Then if $G \supseteq v(\Sigma)$ we have $u(\tau) = \top/G$ for all $\tau \in \Sigma$ so, by the previous version of soundness, $u(\sigma) = \top/G$ hence $v(\sigma) \in G$, as required. $\qquad\square$

Theorem 8.14 *Let X be a set, and let $\pi\colon BT(X) \to$ free(X) be the map induced by $x \mapsto [x]$ and suppose $G \subsetneq$ free(X) is a proper filter. Then*

$$\pi^{-1}G = \{\sigma \in BT(X) : \pi(\sigma) \in G\}$$

is consistent, i.e. $\pi^{-1}G \not\vdash \bot$.

Proof There is a valuation $X \to$ free$(X)/G$ given by sending $x \in X$ to the equivalence class $[x]/G$ of $[x] \in$ free(X) in the quotient algebra free$(X)/G$, and this map sends each $\sigma \in \pi^{-1}G$ to $\top$. Since G is proper, $\top/G \neq \bot/G$

in free$(X)/G$, so this valuation does not send $\bot \in \mathrm{BT}(X)$ to $\bot$ in free$(X)/G$.
Therefore $\pi^{-1}G$ is consistent by soundness. $\qquad\qquad\qquad\qquad\square$

Definition 8.15 Let F be a filter in a boolean algebra B. We say that F is *prime*
if it is proper and whenever $a, b \in B$ with $a \vee b \in F$ then either a or b is in F.

   Dually, an ideal I is *prime* if it is proper and whenever $a, b \in B$ with $a \wedge b \in I$
then either a or b is in I.

Proposition 8.16 *Let B be a boolean algebra and $F \subsetneq B$ a proper filter. Then
the following are equivalent.*

- *F is maximal, i.e. there is no filter G with $F \subsetneq G \subsetneq B$.*
- *F is prime.*
- *F has the property that for all $x \in B$ either $x \in F$ or $x' \in F$.*
- *B/F is isomorphic to the two-element boolean algebra $\{\top, \bot\}$.*

Proof If F is maximal and $a \vee b \in F$ with neither $a, b \in F$ then, by Lemma 8.10,
we may extend F to the filters $G_a = \{g : g \geqslant a \wedge f \text{ for some } f \in F\}$ and $G_b =
\{g : g \geqslant b \wedge f \text{ for some } f \in F\}$ and these contain a, b respectively; hence they
are improper by the maximality of F. Therefore $\bot \geqslant a \wedge f$ and $\bot \geqslant b \wedge g$ for
some $f, g \in F$, so $\bot \geqslant (a \wedge f) \vee (b \wedge g) = (a \vee b) \wedge (f \vee b) \wedge (a \vee g) \wedge (f \vee g) \in
F$, hence F is also not proper, contradicting assumption. So F is maximal
implies that F is prime.

   If F is prime then $x \vee x' = \top \in F$ so one of $x, x' \in F$, so F has the third
property listed in the proposition.

   If F has the property that one of $x, x' \in F$ for each x, then the equivalence
class x/F of x is either F itself (if $x \in F$) or $\{x' : x \in F\}$ (if $x' \in F$) as you may
check. Therefore there are only two such equivalence classes and B/F is the
two-element boolean algebra.

   If B/F is isomorphic to $\{\top, \bot\}$ then $F = \{x \in B : x/F = \top/F\}$ and so for
a filter G with $F \subsetneq G \subseteq B$ we have $x \in G$ for some $x/F = \bot/F$. It follows
from the definition of the equivalence class x/F that $x \vee f = \bot$ for some $f \in F$
i.e. $G = B$. $\qquad\qquad\qquad\qquad\qquad\qquad\qquad\qquad\qquad\qquad\square$

Maximal proper filters are usually called *ultrafilters*. The previous proposition
shows that this notion of ultrafilter coincides exactly with prime filter. The
following theorem stating that ultrafilters exist is an important consequence
of Zorn's Lemma. It is normally known as the Prime Ideal Theorem, and we
shall stick with this convention, despite the fact that we shall state the result
in its equivalent form concerning filters. It is in fact just another form of the
Completeness Theorem for propositional logic.

Theorem 8.17 (Boolean Prime Ideal Theorem) *Let B be a boolean algebra and suppose $F \subsetneq B$ is a proper filter. Then there is an ultrafilter $G \supseteq F$ of B.*

Proof Let $X = B$ considered as a set of letters, and consider the valuation $v: X = \mathrm{BT}(B) \to B$ given by $v(x) = x$. Then the set of statements $\Sigma = v^{-1}(F)$ is consistent by the Soundness Theorem so by the previous version of completeness there is a valuation $w: \mathrm{BT}(X) \to \{\top, \bot\}$ making each $\sigma \in \Sigma$ true. Then $w^{-1}(\top)$ is a maximal filter in B extending F. $\qquad\square$

8.2 Examples and exercises

Exercise 8.18 Let $f: B \to C$ be a homomorphism of boolean algebras and let $F \subseteq C$ be a filter in C. Then $f^{-1}F = \{x \in B : f(x) \in F\}$ is a filter in B. If F is a proper filter in C then $f^{-1}F$ is a proper filter in B.

Exercise 8.19 If I is a prime ideal in B then its complement $B - I$ is a prime filter. Similarly the complement $B - F$ of a prime filter is a prime ideal.

Exercise 8.20 Let B be a finite boolean algebra. Show that B is isomorphic to the boolean algebra $P(X)$ of all subsets of some (finite) set X. (Hint: say an element $a \subset B$ is an *atom* if $\bot < a$ and there is no $x \subset B$ with $\bot < x < a$. Let X be the set of atoms of X and define $h(b) = \{x \subseteq X : x \leqslant b\}$.)

Exercise 8.21 Let X be a finite set. Show that the free algebra free(X) on X is isomorphic to the algebra $P(X)$ of all subsets of X with $\subseteq$, $\cup$, $\cap$. More generally, if X is a set, possibly infinite, show that free(X) is isomorphic to the subalgebra of $P(X)$ consisting of all of those subsets of X which either are finite or else have finite complement.

Exercise 8.22 Let X be a set of proposition letters and $B =$ free(X), the free algebra on X. Suppose that x, y are distinct letters from X. Show that the equivalence relation $\sim$ of Lemma 8.11 does not make x and y equivalent.

Exercise 8.23 Let B be a boolean algebra, and let $X = B$. Form the free algebra $A =$ free(X), and let $v: A \to B$ be the homomorphism induced by sending the proposition letter $b \in X = B$ to itself. (You will need to apply the previous exercise to show this map is well defined.) Let F be the set of $a \in A$ such that $v(a) = \top$. Show that F is a filter and A/F is isomorphic to B.

Exercise 8.24 (For those who have read Section 5.3 on the algebra of Boole.) Let R be a boolean ring and B the boolean algebra corresponding to it. Show that $I \subseteq R$ is an ideal of R in the sense of ring theory (i.e. is non-empty, is closed under $+$ and satisfies $xy \in I$ whenever one of x or y is in R) if and only if I is an ideal of B in the sense of boolean algebras.

8.3 Tychonov's Theorem*

The main theorem of this chapter was the Boolean Prime Ideal Theorem, the fact that every proper filter in a boolean algebra can be extended to an ultrafilter, i.e. a maximal and hence prime proper filter. Ultrafilters have many applications in mathematics, including to logic itself, and to infinitary combinatorics.

One nice application of ultrafilters is their use in proving Tychonov's Theorem, that an arbitrary product of compact topological spaces is compact. We say an *open cover* U_i $(i \in I)$ of a topological space X is a family of open sets whose union is the whole of X, and the *closure* of a set $A \subseteq X$ is the set of $x \in X$ such that every non-empty open U containing x also contains some $a \in A$. The closure of A is denoted $\overline{A}$. We start with a characterisation of *compact* topological spaces.

Proposition 8.25 *Let X be a topological space, and $B = P(X)$ the boolean algebra of subsets of X. Then the following are equivalent.*

- *Any open cover U_i $(i \in I)$ of X has a finite subcover U_{i_n} $(n = 0, 1, \ldots, k)$.*
- *For any proper filter $F \subseteq B$ the intersection $\bigcap \{\overline{A} : A \in F\}$ is non-empty.*

Proof For one direction, suppose $F \subseteq B$ is a filter. Then $U = \{\overline{A}^c : A \in F\}$, the set of complements of closures of $A \in F$, is a collection of open sets. If $\bigcap \{\overline{A} : A \in F\} = \varnothing$ then there is no $x \in X$ in all the $\overline{A}$ $(A \in F)$, so U is an open cover of X. Therefore there are $A_1, \ldots, A_n \in F$ such that $\overline{A_1}^c \cup \ldots \cup \overline{A_n}^c = X$ hence $\overline{A_1} \cap \ldots \cap \overline{A_n} = \varnothing$ hence $A_1 \cap \ldots \cap A_n = \varnothing$, so F is not proper.

For the other direction, let U_i $(i \in I)$ be an open cover of X with no finite subcover. Then the U_i generate a proper filter,

$$F = \{A : A \supseteq U_{i_1}^c \cap \ldots \cap U_{i_n}^c, \text{ some } n, i_1, \ldots, i_n\},$$

and $\bigcap \{\overline{A} : A \in F\} = \bigcap \{U_i^c : i \in I\} = \varnothing$ as the U_i cover X. □

Definition 8.26 A topological space X satisfying either of the conditions in the last proposition is said to be *compact*.

For *metric spaces*, rather than the more general topological spaces considered here, compactness turns out to be equivalent to the idea of sequential compactness discussed earlier in Example 1.15.

Definition 8.27 Let X_i be a topological space for each $i \in I$. The *product space* is the set $X = \prod\{X_i : i \in I\}$ of all functions $f : I \to \bigcup\{X_i : i \in I\}$ such that $f(i) \in X_i$ for all i. This is given the topology in which the open sets of X are the sets which are unions of *basic open* sets of the form $\prod\{U_i : i \in I\}$, where U_i ranges over open subsets of X_i and U_i *must equal X_i for all but finitely many* $i \in I$.

Definition 8.28 The product space $X = \prod\{X_i : i \in I\}$ has *projection functions* $\pi_i : X \to X_i$ defined by $\pi_i(f) = f(i)$, i.e. evaluating the function f at the coordinate i. Similarly a set $A \subseteq X$ has a projection $\pi_i(A) = \{\pi_i(a) : a \in A\} \subseteq X_i$.

Exercise 8.29 The projection functions $\pi_i : X \to X_i$ defined on the product $X = \prod\{X_i : i \in I\}$ are all continuous. In fact, the topology of the product space $X = \prod\{X_i : i \in I\}$ can be characterised as the topology with the minimum of open sets to make these projection functions continuous.

Theorem 8.30 (Tychonov's Theorem) *Let X_i be compact topological spaces for all $i \in I$. Then the product space $X = \prod\{X_i : i \in I\}$ is also compact.*

Proof Let $F \subseteq P(X)$ be a proper filter, and apply the Boolean Prime Ideal Theorem to obtain an ultrafilter $G \supseteq F$. For each $i \in I$ we obtain an ultrafilter G_i of $P(X_i)$ by projection on X_i defined by $G_i = \{\pi_i(A) : A \in G\}$ where $\pi_i(A) = \{\pi_i(x) : x \in A\}$. It is easy to check that G_i is an ultrafilter as claimed. This is because each G_i is non-empty as G is, and if $B \supseteq \pi_i(A)$ then $B = \pi_i(C)$ where $C = \{f : \pi_i(f) \in B\} \supseteq A$. Also if $A, B \in G$ then

$$\pi_i(A) \cap \pi_i(B) \supseteq \{\pi_i(f) : f \in A \cap B\} = \pi_i(A \cap B) \in G_i$$

and clearly $\varnothing \notin G_i$ since $\varnothing \notin G$.

By the compactness of X_i, there is for each i some $a \in X_i$ in the closure of each $\pi_i(A) \in G_i$. Using the Axiom of Choice we select one such $a_i \in X_i$ for each i, and define $a \in X$ to be the function $a : i \mapsto a_i$. Then it remains to check that $a \in \overline{A}$ for each $A \in G$. That is, we need to show that if $U \subseteq X$ is an open neighbourhood of a then $U \cap A \neq \varnothing$. As open sets in X are unions of basic open sets, it suffices to prove this in the case when $U = \prod\{U_i : i \in I\}$ is such a basic open set. Fix some such U and consider an index i. Then as $a_i \in \overline{\pi_i(A)}$

and $a_i \in U_i = \pi_i(U)$ there is some $b_i \in U_i \cap \pi_i(A)$. It follows that the function $b: i \mapsto b_i$ is in $U \cap A$, as required. $\qquad\square$

We remark that the above proof uses the Axiom of Choice in several different places. Firstly there is an implicit application of the Axiom of Choice in the use of the Boolean Prime Ideal Theorem (which needed Zorn's Lemma); and secondly there are two explicit uses at the end to select our a. There is also, rather more subtly, an implicit use of the Axiom of Choice in assuming that the product space X is non-empty in the first place. Tychonov's Theorem is actually equivalent to the Axiom of Choice, and Choice cannot be avoided. On the other hand, the Boolean Prime Ideal Theorem is in fact weaker than the full Axiom of Choice (though it does require at least some form of Choice for its proof).

Exercise 8.31 Let the topological space X be compact and *Hausdorff* (i.e. for each $x \neq y$ in X there are open sets U, V such that $x \in U, y \in V$ and $U \cap V = \varnothing$). Suppose that $F \subseteq P(X)$ is an ultrafilter. Show that $\bigcap \{\overline{A} : A \in F\}$ is a singleton set.

Exercise 8.32 Let X_i be non-empty sets for each $i \in I$. Show that the product $X = \prod \{X_i : i \in I\}$ is non-empty. Conversely, show that the statement you have just proved implies the Axiom of Choice.

8.4 The Stone Representation Theorem*

Exercise 8.20 showed that a finite boolean algebra is isomorphic to the boolean algebra $P(X)$ of all subsets of some (finite) set X. This is a *representation theorem* showing that finite boolean algebras are represented in a nice way. For infinite boolean algebras there is a similar but more complicated representation theorem, called the Stone Representation Theorem. We will explain this result here. In this section we work with a boolean algebra B, which is usually infinite.

Definition 8.33 Given a boolean algebra B, the dual of B, dual of boolean algebra, B^{dual}, is the set of ultrafilters $F \subseteq B$. We give B^{dual} a topology by defining the set T of all open sets. Here, T is defined to be the minimum of sets that makes all collections of filters of the form

$$U_a = \left\{ F \in B^{\text{dual}} : a \in F \right\},$$

i.e. all collections of filters F containing a specific $a \in B$, open. In other words, the set of all U_a forms a base of open sets for the topology (it is closed under intersections because $U_a \cap U_b = U_{a \wedge b}$) and T is defined to be the set of all $U \subseteq B^{\text{dual}}$ which are a union of sets of the form U_a. We call B^{dual} with this topology the *dual space* of B.

Definition 8.34 A topological space X is *totally disconnected* if for all $x, y \in X$ with $x \neq y$ there are open $U, V \subseteq X$ with $x \in U$, $y \in V$ and $X = U \cup V$ is a *disjoint* union of the two sets.

Proposition 8.35 *The topological space B^{dual} is a compact totally disconnected topological space, in which the clopen subsets (i.e. sets which are both closed and open) are exactly the sets of the form U_a for $a \in B$.*

Proof This is an application of the Boolean Prime Ideal Theorem. We start by proving compactness.

Let $\{U_i : i \in I\}$ be an open cover of B^{dual}. We must show that there is a finite subcover. Each U_i is a union of basic open sets, so there is a cover $\{U_a : a \in A\}$ for some $A \subseteq B$ such that each U_a $(a \in A)$ is a subset of some U_i. It suffices therefore to show that there is a finite subcover of $\{U_a : a \in A\}$. We shall assume, to get a contradiction, that there is no such subcover.

Consider the filter

$$F = \left\{ x \in B : x \geqslant a_1' \wedge \ldots \wedge a_k', k \in \mathbb{N}, a_1, \ldots, a_k \in A \right\}.$$

This is clearly a filter. It is also proper since if $\perp \in F$ then $a_1' \wedge \ldots \wedge a_k' = \perp$ for some k and some $a_1, \ldots, a_k \in A$; we show that this would imply that $U_{a_1} \cup \ldots \cup U_{a_k} = B^{\text{dual}}$. If $G \in B^{\text{dual}}$, so $G \subseteq B$ is a (proper) ultrafilter and $G \notin U_{a_i}$ for each i then $a_i \notin G$ for each i (by the definition of U_{a_i}) so $a_i' \in G$ for each i, as G is an ultrafilter. It would follow that $\perp = a_1' \wedge \ldots \wedge a_k' \in G$, which is impossible.

Now, since F is a proper filter of B it extends to an ultrafilter $F \subseteq G \in B^{\text{dual}}$. But by construction $G \notin U_a$ for each $a \in A$, as $a' \in F \subseteq G$ and therefore $a \notin G$, otherwise G would contain $a \wedge a' = \perp$. Thus $\{U_a : a \in A\}$ is not in fact a cover of B^{dual}, and this is our required contradiction completing the proof of compactness.

The other properties now follow more easily. To see that B^{dual} is totally disconnected, note first that $U_a \cup U_{a'} = B^{\text{dual}}$ is a disjoint two-set cover of B^{dual} for each $a \in B$. This is because each ultrafilter G of B contains exactly one of a, a'. Thus each U_a is clopen, and if $F, G \in B^{\text{dual}}$ are distinct there is some $a \in F \setminus G$ or $b \in G \setminus F$; in the first case we have $F \in U_a$ and $G \in U_{a'}$, and in the second case $F \in U_{b'}$ and $G \in U_b$. This shows that B^{dual} is totally disconnected.

Now suppose $U \subseteq B^{\text{dual}}$ is clopen. We want to show $U = U_a$ for some $a \in B$.

As U is open there are U_{a_i} $(i \in I)$ such that $U = \bigcup\{U_{a_i} : i \in I\}$, and as U^c is open there are U_{b_j} $(j \in J)$ such that $U^c = \bigcup\{U_{b_j} : j \in J\}$. Thus the U_{a_i} and U_{b_j} form an open cover of B^{dual}, and by compactness there is a finite subcover. In particular $U = U_{a_1} \cup \ldots \cup U_{a_k}$ for some $a_1, \ldots, a_k \in B$. It follows that $U = U_a$ where $a = a_1 \vee \ldots \vee a_k$, for given an ultrafilter $G \in B^{\text{dual}}$ we have: if $G \in U_a$ then some $a_i \in G$ (for if $a_i \notin G$ for all i then $a_i' \in G$ as G is an ultrafilter, so $\perp = (a_1 \vee \ldots \vee a_k) \vee (a_1' \wedge \ldots \wedge a_k') \in G$ which is impossible) hence $G \in U_{a_i}$ so $G \in U$; and conversely, if $G \in U$ then $a_i \in G$ for some i so $a = a_1 \vee \ldots \vee a_k \geqslant a_i$ is also in G. Thus the clopen sets are precisely the basic open sets. $\qquad\square$

The interesting thing about clopen sets is that finite unions and intersections of clopen sets are clopen, as are complements of clopen sets. In other words, the collection of clopen subsets of a topological space forms a boolean algebra with the usual $\subseteq$ relation, and $\cup, \cap$. This now gives the promised representation theorem.

Theorem 8.36 (Stone Representation Theorem) *Let B be a boolean algebra and B^{dual} its dual space. Then the map*

$$a \mapsto U_a$$

is an isomorphism from the boolean algebra B to the boolean algebra of clopen subsets of B^{dual}.

Proof Most of the ideas have already been presented in the last proposition, and we only need to tie some loose ends. In particular, the map $a \mapsto U_a$ is a homomorphism of boolean algebras from B to the algebra of clopen sets in B^{dual}.

To see $U_{a \wedge b} = U_a \cap U_b$, take an ultrafilter G and observe $a \wedge b \in G$ implies $a \in G$ and $b \in G$ as $a, b \geqslant a \wedge b$ and conversely $a \in G$ and $b \in G$ implies $a \wedge b \in G$ as G is closed under $\wedge$.

To see $U_{a \vee b} = U_a \cup U_b$, observe $a \vee b \in G$ implies $a \in G$ or $b \in G$ as G is an ultrafilter and the alternative would be $a' \in G$ and $b' \in G$. Conversely $a \in G$ or $b \in G$ implies $a \vee b \in G$ as $a \vee b \geqslant a, b$.

That $U_{a'} = U_a{}^c$ has already been noted, as $U_a \cup U_{a'} = B^{\text{dual}}$ is a disjoint union. Also $U_\perp = \varnothing$ and $U_\top = B^{\text{dual}}$ as every ultrafilter G contains $\top$, and none contains $\perp$. This shows $a \mapsto U_a$ is a homomorphism of boolean algebras.

Finally, this homomorphism is onto, since every clopen set is U_a for some a, and one-to-one since if $a \neq b$ then $F = \{x : x \geqslant a \wedge b'\}$ is a proper filter (proper because $a \wedge b' = \perp$ implies $a = b$) and so is contained in an ultrafilter $G \in U_a \setminus U_b$. $\qquad\square$

If we now take one step back we may see that this correspondence between the boolean algebra and its dual space is much deeper and more powerful.

Definition 8.37 Say a topological space X is a *Stone space* if it is compact and totally disconnected.

Let BOOL denote the class of all boolean algebras and STONE the class of all Stone spaces. Then the Stone Representation Theorem uses two important maps, D: BOOL $\to$ STONE, taking B to $S(B) = B^{\text{dual}}$, and E: STONE $\to$ BOOL, taking X to the boolean algebra $B(X) = X^{\text{clopen}}$ of clopen subsets of X. The Stone Theorem says that $B(S(B))$ is isomorphic to B. In fact a similar result holds the other way round too, and the maps D and E are inverse to each other.

Theorem 8.38 *Let X be a Stone space, $B = X^{\text{clopen}}$ the boolean algebra of clopen subsets of X, and $Y = B^{\text{dual}}$ the dual space of B. Then X and Y are homeomorphic.*

Proof We must define a map $X \to Y$ and show it is one-to-one, onto, continuous, and with continuous inverse. Given $x \in X$ we define

$$D_x = \{U \subseteq X : U \text{ clopen}, x \in U\}.$$

The set D_x is a set of clopen subsets of X, hence a subset of B and our map will be $x \mapsto D_x$. For this map to be well defined as a map $X \to Y$ we need to check that D_x is indeed an ultrafilter. But if $U, V \subset X$ are any clopen sets and $x \in U$ and $x \in V$ then $x \in U \cap V$ and $U \cap V$ is clopen. Similarly if $x \in U \subseteq V$ then $x \in V$. Thus D_x is a filter. It is proper as it does not contain $\varnothing$ and an ultrafilter because if U is any clopen set then U^c is clopen and x is in exactly one of U, U^c, hence exactly one of U, U^c is in D_x.

To check that $x \mapsto D_x$ is one-to-one, suppose $x \neq y$ are elements of X. Then by total disconnectedness there are disjoint clopen $U, V \subseteq X$ such that $x \in U$, $y \in V$ and $X = U \cap V$. So $U \in D_x$ and $V \in D_y$ and $D_x \neq D_y$.

For the onto property, we use compactness. Let $G \subseteq B = X^{\text{clopen}}$ be an ultrafilter. Then by compactness and Proposition 8.25 there is some x in the intersection $\bigcap \{\overline{U} : U \in G\}$. As each $U \in G$ is already closed, we may write this more simply as $x \in \bigcap G$. Now let $U \subseteq X$ be clopen. Then $X = U \cup U^c$ is a disjoint union of clopen sets and G contains exactly one of U, U^c as it is an ultrafilter. Since $x \in \bigcap G$, G must contain whichever of U, U^c that contains x, and hence $G = D_x$.

For continuity, suppose $U \subseteq Y$ is basic open, of the form $U = U_V$, where

$V \in B$, i.e. $V \subseteq X$ is clopen. Then $U = \{D_x : V \in D_x\}$ and the inverse image of U under $x \mapsto D_x$ is just V, which is open. Similarly, every open set of a totally disconnected space is a union of clopen sets. If $V \subseteq X$ is clopen then the image of V under $x \mapsto D_x$ is $\{D_x : x \in V\}$ which equals $\{D_x : V \in D_x\}$, a clopen set in Y. □

For our final refinement of these ideas, we can consider the classes BOOL and STONE together with their familiar notion of 'homomorphisms'. (In other words, we consider BOOL and STONE as categories, though we shall not in fact need or use any terminology from category theory in this book.) For $A, B \in$ BOOL, the notion of 'homomorphism' is that of ordinary homomorphisms $A \to B$ of boolean algebras. For $X, Y \in$ STONE the corresponding notion of 'homomorphism' is that of a *continuous map* $X \to Y$. Then, not only do boolean algebras and their Stone-space duals correspond, but so also do homomorphisms between them – except that in the correspondence of homomorphisms the direction of the maps must be reversed. (It is this reversal of the direction of maps that is characteristic of 'duality' and the reason for calling B^{dual} the 'dual' of B in the first place.) Specifically, we have the following theorem.

Theorem 8.39 *(a) Let $A, B \in$ BOOL be boolean algebras and let $h: A \to B$ be a homomorphism of boolean algebras. Then there is a continuous map $S(h): S(B) \to S(A)$ from the dual $S(B) = B^{\text{dual}}$ of B to the dual $S(A) = A^{\text{dual}}$ of A given by $a \in S(h)(y)$ if and only if $h(a) \in y$. Furthermore, if h is one-to-one then $S(h)$ is onto, and if h is onto then $S(h)$ is one-to-one.*

(b) Let $X, Y \in$ STONE be Stone spaces and $\theta: X \to Y$ a continuous map. Then there is a homomorphism $B(\theta): B(Y) \to B(X)$ from the boolean algebra $B(Y) = Y^{\text{clopen}}$ of clopen sets in Y to the boolean algebra $B(X) = X^{\text{clopen}}$ of clopen sets in X given by $a \in \theta(y)$ if and only if $B(\theta)(a) \in y$. Furthermore, if θ is one-to-one then $B(\theta)$ is onto, and if θ is onto then $B(\theta)$ is one-to-one.

Proof (a) Given an ultrafilter $G \subseteq B$ define $\theta(G) = \{a \in A : h(a) \in G\}$. This is a proper filter because: $\perp \notin \theta(G)$ since $h(\perp) = \perp \notin G$; if $a, b \in \theta(G)$ then $h(a), h(b) \in G$ so $h(a \wedge b) = h(a) \wedge h(b) \in G$ so $a \wedge b \in \theta(G)$; and if $a \geqslant b \in \theta(G)$ then $h(a) \geqslant h(b) \in G$ so $h(a) \in G$ and $a \in \theta(G)$. It is an ultrafilter because, given a, either $h(a) \in G$ or $h(a') = h(a)' \in G$. $\theta = S(h)$ is the required map $S(B) \to S(A)$.

To check θ is continuous consider a basic open $U_a = \{H \subseteq A : a \in H\} \subseteq A^{\text{dual}}$. Then $\theta(G) \in U_a$ if and only if $a \in \theta(G)$, which holds if and only if $h(a) \in G$, so $\theta^{-1}U_a = U_{h(a)}$ is open.

Given $H \subseteq A$ let $G = \{h(a) : a \in H\}$ so $H \subseteq \{a \in A : h(a) \in G\}$, and if h is one-to-one then for each $a \in H$ there is a unique $b \in G$ such that $b = h(a)$, hence θ is onto. If h is onto and G_1, G_2 are such that $\theta(G_1) = \{a \in A : h(a) \in G_1\} = \{a \in A : h(a) \in G_2\} = \theta(G_2)$ then for each $b \in G_1$ there is $a \in A$ with $b = h(a)$ so $a \in \theta(G_1) = \theta(G_2)$ hence $b = h(a) \in G_2$. Thus by symmetry $G_1 = G_2$ and θ is one-to-one.

(b) Given $X, Y \in$ STONE and continuous $\theta : X \to Y$ we let $h(U)$ be the set $\{x \in X : \theta(x) \in U\}$ for a clopen $U \subseteq Y$. This is clopen in X by the continuity of θ and a similar amount of straightforward axiom-checking shows that h is a homomorphism of boolean algebras, is one-to-one if θ is onto, and is onto if θ is one-to-one. Thus $B(\theta) = h$ is the required map. $\qquad\square$

9

First-order logic

9.1 First-order languages

Propositional logic is the logic of statements that can be true or false, or take some value in a boolean algebra. The logic of most mathematical arguments involves more than just this: it involves mathematical objects from one or other domain, such as the set of natural numbers, real numbers, complex numbers, etc. If we introduce such objects into our formal system for proof we get what is known as first-order logic, or predicate logic.

As for any of our other logics, first-order logic would not be so interesting if it was just a system for writing and mechanically checking formal proofs for one particular domain of mathematical work. But fortunately it can be interpreted in a rather general class of mathematical structures and the theory of these structures is a sort of generalised algebraic theory that applies equally well to groups, rings, fields, and many other familiar structures, so first-order logic can be applied to a wide range of mathematical subject areas.

There are Completeness and Soundness Theorems for first-order logic too. In a similar way to the Completeness and Soundness Theorems we have already seen, they can be read as stating the correctness and adequacy of our logical system, or as much more interesting constructive statements that enable new structures to be created and analysed.

We will start here by discussing the idea of first-order language, and the sorts of things that can (and cannot) be expressed in first-order logic. Later on, we will give some rules for a proof system for this logic. The rules we give will be precise versions of logical manoeuvres familiar from many informal arguments or proofs, and will build on the proof system for the propositional calculus given earlier.

Consider for the moment building a theory of the reals $\mathbb{R}$. We will want to talk about specific real numbers, so will need variables to represent reals.

We also require symbols for special real numbers such as 0 or 1. We will want to say when two real numbers are equal, so we will need the symbol = for equality. We also need to combine real numbers with familiar functions such as $x \mapsto -x$ and $(x, y) \mapsto x + y$, etc., so we will need symbols to represent these. We may need to compare two real numbers and determine which is the greater. A symbol for the < relation is necessary here. Finally, we need two special symbols $\forall$ and $\exists$ (called *quantifiers*) to represent the phrases 'for all ...' and 'there exists ...' common in mathematical arguments. If we put all of this together we have a first-order language for the real numbers. There are similar first-order languages for groups, posets, boolean algebras, etc. The next definition brings together all of these into one general framework.

Definition 9.1 A *first-order language* consists of the following symbols:

- $\wedge$, $\vee$, $\neg$, $\top$, $\bot$ for propositional logic;
- an infinite set of variables, $x, y, z, \ldots$;
- the symbols '$=$', '$\forall$', '$\exists$';
- a (possibly empty) set of constant symbols, such as 0, 1;
- a (possibly empty) set of function symbols, such as $+$, $\times$, $-$;
- a (possible empty) set of relation symbols, such as <;
- the punctuation symbols '(', ')' and ', '.

The *logical symbol*s of a first-order language are $\wedge$, $\vee$, $\neg$, $\top$, $\bot$, $=$, $\forall$, $\exists$; all the first-order languages we consider have all of these as well as variables and punctuation symbols. (Countably infinitely many variables is always sufficient.) The remaining symbols are special to the particular language and are called *non-logical symbol*s. Thus we can specify a first-order language by giving the constant, relation and function symbols and taking the other symbols for granted. For example we may discuss 'the first-order language for the reals with $0, 1, +, \times, -, <$'.

Of course, a language is more than just a collection of symbols. We need to say how these symbols are combined and what the resulting strings of symbols (also called expressions or statements) mean. From the point of view of experts (or pedants) my explanation of how this is done will be very informal, relegating the more formal details to a discussion elsewhere. For beginners, the discussion that I do give will be quite complicated enough, especially as we are working in sufficient generality to apply the ideas to such a wide range of mathematical topics. Fortunately, the syntactical constructions and their intended meanings by and large follow normal mathematical usage anyway. My advice to a beginner is to read and write these expressions in as natural a way as possible, and the likelihood is that you will be right.

We start by defining terms, the expressions representing a mathematical object, such as a number. One problem is that we want to use our non-logical function symbols to form terms, but some, like $^{-1}$ for reciprocal, take only one argument (as in x^{-1}), others such as $+$ for addition take two, and others take more. We insist that every non-logical function symbol f in a first-order language takes a fixed number of arguments. This number will be denoted n_f and is called the *arity* of f. Strictly speaking, the arities of all non-logical function symbols have to be specified when a first-order language is defined, though in most cases the arities will be clear, as for addition when the arity of $+$ is conventionally two. A function symbol of arity one is said to be *unary*, one of arity two is a *binary* function symbol, and one of arity three is *ternary*.

Definition 9.2 A *term* in a first-order language is an expression or string of symbols that represents a mathematical object, such as a number.

- Each constant symbol, such as 0 or 1 is a term.
- Each variable from $x, y, z, \ldots$ is a term.
- If f is a function symbol and $t_1, \ldots, t_n$ are n terms where $n = n_f$ is the arity of f then $f(t_1, \ldots, t_n)$ is also a term.

The terms of a first-order language are the finite expressions that can be built by finitely many applications of the above rules only.

Often, we prefer to write terms in a more natural way as $(a + b)$ rather than $+(a, b)$, say, and omit brackets when there is no danger of confusion.

It is not always clear exactly which number a term such as $(x + 1)$ represents. After all we have not said which numbers are represented by the variables $x, y, z, \ldots$. On the other hand, terms that do not involve variables are more definite. These terms are said to be closed.

Definition 9.3 A *closed term* is a term which does not involve any variable symbols.

All of these rules are like the rules of a game on a set of symbols: they can be given accurately without talking about meanings or semantics, but it helps to have an idea of the meanings we will eventually choose as we go along.

Now that we have defined terms, we attempt to define the statements of a first-order language. To do this we need the idea of the *arity* of each relation symbol. Like that for functions, this is the number of arguments it receives. For example $<$ has arity two, as does the *logical* relation symbol $=$ since we use these symbols to compare two numbers, as in $1 < 2$ or $x = y$. The relation Odd, used in 'Odd(x)' to indicate that x is an odd number, has arity one or in other words is *unary*. We also say that a relation symbol of arity two is *binary*, a relation symbol of arity three is *ternary*, etc. Every relation symbol R of a first-order language has a fixed arity, denoted n_R and these arities all have to be specified when the language is defined, though where possible we follow common usage for standard relations like $<$, $\sim$, etc.

Definition 9.4 A *statement* or *formula* of a first-order language is a finite expression or string of symbols built using the following rules only.

- If t, s are terms of the language then $(t = s)$ is a formula.
- If R is a relation symbol of the language of arity $n = n_R$, and $t_1, \ldots, t_n$ are terms of the language then $R(t_1, \ldots, t_n)$ is a formula.
- If ϕ, ψ are both formulas then so are $(\phi \lor \psi)$, $(\phi \land \psi)$, $\neg \phi$, $\top$ and $\bot$.
- If ϕ is a formula and x is a variable then $\forall x \, \phi$ and $\exists x \, \phi$ are formulas.

Statements of the form $(t = s)$ or $R(t_1, \ldots, t_n)$ are called *atomic formulas* as other statements are built from these.

As before, we may choose to write some formulas in a more natural way, as $(x < y)$ rather than $<(x, y)$, say, and omit brackets where there is no danger of confusion. We typically use lower case Greek letters to refer to formulas.

According to the definition above, a formula θ may, and usually will, contain substrings which are other formulas. These are called *subformulas* of θ.

In a way exactly analogous to the case of terms, the meaning of formulas may or may not depend on the meaning of particular variables, but the rules are more complicated because of the quantifiers $\forall x \, \phi$ and $\exists x \, \phi$. To understand the rules for variables and for quantifiers you will need to keep in mind the picture of how a formula is built from subformulas.

Definition 9.5 The *scope of a quantifier* $\forall x \ldots$ that occurs in a formula θ is the subformula of θ consisting of the quantifier itself and the subformula immediately following this quantifier. So if θ is $\ldots \forall x \, \phi \ldots$ where ϕ is a subformula of θ then the scope of $\forall x \ldots$ is the subformula $\forall x \, \phi$.

Similarly, the *scope of the quantifier* $\exists x \ldots$ in a formula θ is the subformula $\exists x \, \phi$ of θ consisting of the quantifier and the subformula immediately following it.

The bracketing rules for formulas ensure that the scope of a quantifier is uniquely determined: there is always exactly one subformula immediately following any quantifier. (If this is not the case for your formula then you must have accidently omitted some brackets.) The idea is that the scope of the quantifier is the part of the formula for which the variable's meaning is modified by the quantifier.

Definition 9.6 A formula σ is *closed* or is a *sentence* if every occurrence of every variable x in it is in the scope of a matching quantifier $\forall x \ldots$ or $\exists x \ldots$. An occurrence of a variable x in a formula θ is *free* if it is not in the scope of any matching quantifier $\forall x \ldots$ or $\exists x \ldots$; if otherwise then we say this occurrence of x is *bound*.

Example 9.7 Consider for example the first-order language for the reals where the non-logical symbols are constants 0, 1, functions $+$, $\times$, $-$ of arities 2, 2 and 1 respectively, and relation symbol $<$ of arity 2, as well as the logical symbols including $=$. The term $(1 + 1)$ is closed and always represents 2. (Note that 2 is not in our language. This term shows we do not really need it.) On the other hand, the terms $(x + y)$ and $(x \times x)$ are not closed. The first ranges over all real numbers, whereas the second ranges over all non-negative real numbers.

Many familiar statements about real numbers can now be represented. For example $(0 < x) \vee (0 = x)$ is a formula expressing the statement that x is non-negative. (This formula is not closed: it has the variable x occurring free in it.) The statement that square numbers are always non-negative can be expressed by

$$\forall x ((0 < x \times x) \vee (0 = x \times x)).$$

This statement is closed: every occurence of x in it is bound.

The statement that the polynomial $x^2 + x + 1$ has a root is the statement

$$\exists x ((((x \times x) + x) + 1) = 0)$$

and we follow common notation to simplify this to $\exists x (x \times x + x + 1 = 0)$ when it is clear that this should be regarded as an abbreviation for the first. Note that we are only discussing what can or cannot be expressed: the statement just given is actually false for the real numbers!

For a more complicated example, consider the statement that every polynomial of degree 3 has a root. This is expressed as the sentence

$$\forall a \, \forall b \, \forall c \, \forall d \, \exists x (a \times x \times x \times x + b \times x \times x + c \times x + d = 0)$$

which is strictly speaking an abbreviation for

$$\forall a \, \forall b \, \forall c \, \forall d \, \exists x (((((((a \times x) \times x) \times x) + ((b \times x) \times x)) + (c \times x)) + d) = 0).$$

Given a first-order language, there are usually a number of structures in which we can interpret sentences. For example, corresponding to the first-order language with $0, 1, <, +, \times, -$ we have the algebraic structure of the reals as an ordered field. The same language can be interpreted in other structures too, such as the ordered field of the rationals, or the ring of integers, or something completely different. So (as in our chapter on the logic of posets) we can have many different structures interpreting the same sentences – possibly making them true, possibly false. The common features of all such structures are that they all contain a non-empty set of mathematical objects or numbers, called the domain of the structure, they contain elements of the domain interpreting the constant symbols, relations on the domain interpreting the relation symbols, and functions on the domain corresponding to the function symbols. We will return to these structures later after we have given a system of proof for first-order logic.

To define our formal system of proof, we need to use variables and the idea of scope, and there are a number of technical details that should be addressed, but are best omitted or read quickly at first reading. (If the first-order language is used in a natural or sensible way these issues need not crop up.) So we shall introduce some informal notation for formulas with free variables next that covers all the usual situations.

Given a formula θ with free variable x we can regard θ as expressing some property that x might have. We shall write this formula more suggestively as $\theta(x)$. This notation only makes sense however if x is the *only* free variable in θ. If the property $\theta(x)$ also depends on the value y we will eventually make some error, and we want to avoid this.

More generally, we can write a formula θ as $\theta(x_1, \ldots, x_k)$ but only if the free variables in θ are amongst $x_1, \ldots, x_k$. Actually it will not matter much if we list *more* variables than actually occur free, but it certainly will matter if we miss any out. So, from now on when we write 'the formula $\theta(x_1, \ldots, x_k)$' we are stating implicitly that 'the free variables in θ are amongst $x_1, \ldots, x_k$'.

Note that in a formula such as $\theta(x)$, the free variable x may occur more than once in $\theta(x)$. Sometimes we might like to specify these separate occurrences, but there is no good notation for this – we will discuss this point further when needed.

If $\theta(x)$ is a formula involving some free instance or instances of a variable x then $\theta(t)$ denotes the result of replacing all of these instances of x by the term t. At the level of symbols in a string, the operation being carried out here is a substitution of a string t for one or more specified occurrences of a variable symbol x. Similarly, if $\theta(x_1, \ldots, x_k)$ has free instances of variables $x_1, \ldots, x_k$ then $\theta(t_1, \ldots, t_k)$ denotes the result of replacing all these instances with $t_1, \ldots, t_k$ respectively. This is a simultaneous substitution of terms for variables. (The terms $t_1, \ldots, t_k$ might themselves involve variables, so performing n substitutions one after the other may not have the same effect.)

With this idea in mind, we can introduce a formal proof system for first-order logic that extends the proof system for propositional logic.

Definition 9.8 Let L be a first-order language, Σ a set of formulas of L and τ another formula of L. We define $\Sigma \vdash \tau$, read 'Σ *proves* τ' or '*there is a formal proof of* τ *from* Σ', to mean that there is a finite derivation using the following rules.

- (Given Statements Rule) For any $\sigma \in \Sigma$, $\Sigma \vdash \sigma$.

- (Propositional Logic) Any of the rules in the proof system for propositional logic can be used.

- (Equality Rules) (Reflexivity) If t is any term then $\Sigma \vdash (t = t)$; (Symmetry) if t, s are terms and $\Sigma \vdash (t = s)$ then $\Sigma \vdash (s = t)$; (Transitivity) if t, s, r are terms and $\Sigma \vdash (t = s)$ and $\Sigma \vdash (s = r)$ hold then $\Sigma \vdash (t = r)$.

- (Substitution Rule) Given terms $t_1, \ldots, t_k$ and $s_1, \ldots, s_k$ such that we have $\Sigma \vdash \theta(t_1, \ldots, t_k)$ and also for each $i = 1, \ldots, k$ we have $\Sigma \vdash (t_i = s_i)$, then $\Sigma \vdash \theta(s_1, \ldots, s_k)$, provided this substitution is valid.

- ($\exists$-Elimination) To show $\Sigma \cup \{\exists x \, \sigma(x)\} \vdash \theta$ it suffices to show $\Sigma \cup \{\sigma(a)\} \vdash \theta$, provided the substitution is valid and the variable a is a new variable not already free in some formula in Σ nor in θ.

- ($\exists$-Introduction) For any variable x, from $\Sigma \vdash \theta(t)$ you may deduce $\Sigma \vdash \exists x \, \theta(x)$, provided the substitution of the variable x for the term t in θ is valid.

- ($\forall$-Elimination) For any term t, from $\Sigma \vdash \forall x \, \theta(x)$ you may deduce $\Sigma \vdash \theta(t)$, provided the substitution of the term t for x in θ is valid.

- ($\forall$-Introduction) If $\Sigma \vdash \theta(x)$ where no assumption from Σ in the proof involves the free variable x then $\Sigma \vdash \forall x \, \theta(x)$.

A formal proof of τ from Σ is a sequence of formulas ending with τ that shows that $\Sigma \vdash \tau$. Each step in the proof should correspond to one of the proof rules above. Subproofs may be used to indicate where certain assumptions and variables are introduced. We shall write formal proofs as before with vertical lines indicating subproofs.

We usually write the $\exists$-Elimination and $\forall$-Introduction Rules by making a subproof. Then an instance of $\forall$-Introduction looks like the following.

Formal proof

Let x be arbitrary	(1)	
...		
$\theta(x)$	(2)	
$\forall x\, \theta(x)$	(3)	$\forall$-Introduction

An instance of $\exists$-Elimination looks like the following.

Formal proof

$\exists x\, \sigma(x)$	(1)	
Let a satisfy $\sigma(a)$	(2)	
...		
θ	(3)	
θ	(4)	$\exists$-Elimination

In the $\exists$-Elimination Rule one must introduce a variable, which must be a new variable not previously introduced at that point, and a subproof, which must be closed by making a conclusion that does not mention the new variable just introduced.

The next example explains the mysterious phrase, 'provided the substitution of the variable x for t in θ is valid'.

Example 9.9 If $\theta(t)$ is $\exists x\,(x = (t+1))$ (a perfectly reasonable statement which is usually true) then the substitution of x for t is $\exists x\,(x = (x+1))$. This statement is usually false, and the substitution is not valid because in making it, a variable x is introduced into the scope of an already existing quantifier $\exists x$....

A substitution like this would also be invalid if the quantifier were $\forall$ rather than $\exists$. Similarly, substituting the term $(x+y)$ for t would give the formula $\exists x\,(x = ((x+y)+1))$, and this substitution is invalid for the same reason. On the other hand, substituting the term $(y+z)$ for t would give the formula $\exists x\,(x = ((y+z)+1))$, and this substitution is valid as there is no quantifier for y or z.

Problems like that in Example 9.9 are not unique to logical formulas. Care is needed in similar examples in other areas of mathematics. For example, a function $f(t)$ may be defined from another function $g(x)$ by an integral, $f(t) = \int_{-\infty}^{\infty} g(x)\sin(tx)\,dx$. It would be incorrect to substitute a term such as $(x+y)$ for t in the formula for $f(t)$, since this would give the incorrect $\int_{-\infty}^{\infty} g(x)\sin(x(x+y))\,dx$. Instead the bound variable in the integral, x, should be changed first to something different: $f(x+y) = \int_{-\infty}^{\infty} g(s)\sin((x+y)s)\,ds$. We will see later how such a change of variable can also be achieved in the proof system we are studying.

The condition that the variable introduced in the ∃-Elimination Rule is a 'new' one and that no conclusions involving this variable are passed on when the subproof is closed is also very important. In the following, this rule is broken and the newly introduced variable 'leaks out' with disastrous effects. (The assumptions are reasonable but the conclusion clearly is not.)

Example 9.10 Something goes badly wrong in the following incorrect proof.

Formal proof

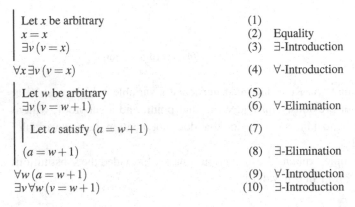

Let x be arbitrary	(1)	
$x = x$	(2)	Equality
$\exists v\,(v = x)$	(3)	∃-Introduction
$\forall x\,\exists v\,(v = x)$	(4)	∀-Introduction
Let w be arbitrary	(5)	
$\exists v\,(v = w+1)$	(6)	∀-Elimination
Let a satisfy $(a = w+1)$	(7)	
$(a = w+1)$	(8)	∃-Elimination
$\forall w\,(a = w+1)$	(9)	∀-Introduction
$\exists v\,\forall w\,(v = w+1)$	(10)	∃-Introduction

In examples like this where a proof is clearly wrong it should always be possible to point out the exact line where one of the rules above has been broken. Conversely, as with any other system of formal proof, any step in a proof should be checkable to see that it obeys the rules exactly. In this case it is line 8 that is erroneous as the variable a still appears free in this line, breaking the rule of ∃-Elimination.

Remark 9.11 Many people like to have a symbol for 'implies'. There are several options for incorporating implies into the system, all of them equivalent. One can add the symbol $\rightarrow$ to the language, modifying the definition of the word 'formula' appropriately, and add the following proof rules.

- ($\rightarrow$-Introduction) If τ can be deduced from $\Sigma \cup \{\sigma\}$ then $(\sigma \rightarrow \tau)$ can be deduced from Σ in one further step.
- ($\rightarrow$-Elimination, also called *modus ponens*) If $(\sigma \rightarrow \tau)$ and σ can be deduced from Σ then τ can be deduced from Σ in one further step.

See Exercise 6.22. Alternatively (and Exercise 6.22 asks you to show that this really is equivalent) you may pretend that there is actually no extra symbol for 'implies' at all, but that 'α implies β' or '$\alpha \to \beta$' is actually just an abbreviation for $(\neg \alpha \vee \beta)$, and use the ordinary rules for $\vee$.

If the set on the left hand side of $\Sigma \vdash \sigma$ is finite, we shall omit the curly brackets and just list the elements of Σ, as in '$\phi, \psi \vdash \sigma$', etc. We will do the same for the other turnstile symbol $\vDash$ when it is introduced. Also, we may occasionally want to put a set Σ on the right hand side of a turnstile symbol, as in $\phi \vdash \Sigma$. This will mean $\phi \vdash \sigma$ for every $\sigma \in \Sigma$.

To learn how these new proof rules work, and in particular understand the idea of quantifiers, we will first look at some example proofs showing equivalences of statements involving quantifiers.

Example 9.12 Let $\theta(x)$ be a formula in which x is free. Then we have the following: $\neg \forall x\, \theta(x) \vdash \exists x \neg \theta(x)$.

Formal proof

$\neg \forall x\, \theta(x)$	(1)	Given
$\exists x \neg \theta(x)$	(2)	Assumption
Let x be arbitrary	(3)	
$\neg \theta(x)$	(4)	Assumption
$\exists x \neg \theta(x)$	(5)	$\exists$-Introduction
$\perp$	(6)	
$\neg \neg \theta(x)$	(7)	RAA
$\theta(x)$	(8)	$\neg$-Elimination
$\forall x\, \theta(x)$	(9)	$\forall$-Introduction
$\perp$	(10)	
$\neg \neg \exists x \neg \theta(x)$	(11)	RAA
$\exists x \neg \theta(x)$	(12)	$\neg$-Elimination

Exercise 9.13 Prove that $\neg \exists x\, \theta(x) \vdash \forall x \neg \theta(x)$. (Use $\exists$-Introduction and $\forall$-Introduction.)

Example 9.14 Let $\theta(x)$ be a formula in which x is free. Then we have the following: $\forall x \neg \theta(x) \vdash \neg \exists x\, \theta(x)$.

Formal proof

$\forall x \neg \theta(x)$	(1)	Given
$\exists x \, \theta(x)$	(2)	Assumption
Let a satisfy $\theta(a)$	(3)	
$\neg\theta(a)$	(4)	$\forall$-Elimination
$\bot$	(5)	
$\bot$	(6)	$\exists$-Elimination
$\neg\exists x \, \theta(x)$	(7)	RAA

Exercise 9.15 Prove that $\exists x \neg \theta(x) \vdash \neg \forall x \, \theta(x)$. (Use $\exists$-Elimination and $\forall$-Elimination.)

The last four examples and exercises show that $\neg \forall x \neg \theta(x)$ and $\exists x \, \theta(x)$ are equivalent, and similarly $\neg \exists x \neg \theta(x)$ and $\forall x \, \theta(x)$ are equivalent. In other words, one of these quantifiers can be defined in terms of the other. Some authors use just one quantifier in their formal language and regard the other as an abbreviation. I feel it is more natural to use both sets of quantifier rules. What is more, since there is an exact symmetry between the two quantifiers, it is difficult to decide which should be dropped and which preserved.

The next example uses unary relation symbols 'P' and 'R' and the $\rightarrow$ symbol. You can read '$P(x)$' as 'x is a Pope' and '$R(x)$' as 'x is in this room', so the example says that if there are two people in this room and at most one Pope then one person in this room is not a Pope.

Example 9.16 There is a proof of $\exists x \, (R(x) \wedge \neg P(x))$ from the two statements $\exists x \, \exists y \, (R(x) \wedge R(y) \wedge \neg(x = y))$ and $\forall x \, \forall y \, (P(x) \wedge P(y) \rightarrow (x = y))$.

First, we write down all the given statements and the negation of the statement we are trying to prove and then apply the $\exists$-Elimination Rule as far as possible.

Formal proof

$\exists x \, \exists y \, (R(x) \wedge R(y) \wedge \neg(x = y))$	(1)	Given
$\forall x \, \forall y \, (P(x) \wedge P(y) \rightarrow (x = y))$	(2)	Given
$\neg\exists x \, (R(x) \wedge \neg P(x))$	(3)	Assumption
Let a satisfy $\exists y \, (R(a) \wedge R(y) \wedge \neg(a = y))$	(4)	
Let b satisfy $R(a) \wedge R(b) \wedge \neg(a = b)$	(5)	
...		

The rest of the proof involves propositional logic and the ∃-Introduction and ∀-Elimination Rules.

Formal proof (continued)

...		
$R(a)$	(6)	∧-Elimination
$R(b)$	(7)	∧-Elimination
$\neg(a = b)$	(8)	∧-Elimination
$\neg P(a)$	(9)	Assumption
$R(a) \wedge \neg P(a)$	(10)	∧-Introduction
$\exists x\,(R(x) \wedge \neg P(x))$	(11)	∃-Introduction
$\bot$	(12)	
$\neg\neg P(a)$	(13)	RAA
$P(a)$	(14)	
$\neg P(b)$	(15)	Assumption
$R(b) \wedge \neg P(b)$	(16)	∧-Introduction
$\exists x\,(R(x) \wedge \neg P(x))$	(17)	∃-Introduction
$\bot$	(18)	
$\neg\neg P(b)$	(19)	RAA
$P(b)$	(20)	
$P(a) \wedge P(b)$	(21)	∧-Introduction
$\forall y\,(P(a) \wedge P(y) \rightarrow (a = y))$	(22)	∀-Elimination
$P(a) \wedge P(b) \rightarrow (a = b)$	(23)	∀-Elimination
$(a = b)$	(24)	→-Elimination
$\bot$	(25)	
$\bot$	(26)	∃-Elimination
$\bot$	(27)	∃-Elimination
$\neg\neg\exists x\,(R(x) \wedge \neg P(x))$	(28)	RAA
$\exists x\,(R(x) \wedge \neg P(x))$	(29)	

This example used the equality symbol, but did not need any of the special rules for equality, such as the Substitution Rule. A variation of our 'Pope' example says that if there is at most one Pope, and at least one Pope that is not in this room, then everyone in this room is not a Pope. This requires the Substitution Rule. The overall strategy for the proof is clear: using ∀-Introduction and →-Introduction we need to show that the given statements together with $P(x)$ imply $\neg R(x)$.

Example 9.17 There is a proof of $\forall x\,(R(x) \rightarrow \neg P(x))$ from the two statements $\forall x\,\forall y\,(P(x) \wedge P(y) \rightarrow (x = y))$ and $\exists x\,(P(x) \wedge \neg R(x))$.

Formal proof

$\forall x \forall y\,(P(x) \wedge P(y) \to (x = y))$	(1)	Given
$\exists x\,(P(x) \wedge \neg R(x))$	(2)	Given
Let x be arbitrary	(3)	
$\quad R(x)$	(4)	Assumption
$\quad\quad$ Let a satisfy $P(a) \wedge \neg R(a)$	(5)	
$\quad\quad P(a)$	(6)	$\wedge$-Elimination
$\quad\quad \neg R(a)$	(7)	$\wedge$-Elimination
$\quad\quad\quad P(x)$	(8)	Assumption
$\quad\quad\quad (P(x) \wedge P(a))$	(9)	$\wedge$-Introduction
$\quad\quad\quad \forall y\,(P(a) \wedge P(y) \to (a = y))$	(10)	$\forall$-Elimination
$\quad\quad\quad (P(a) \wedge P(x) \to (a = x))$	(11)	$\forall$-Elimination
$\quad\quad\quad (a = x)$	(12)	$\to$-Elimination
$\quad\quad\quad \neg R(x)$	(13)	Substitution, lines 7 and 12
$\quad\quad\quad \perp$	(14)	
$\quad\quad \neg P(x)$	(15)	RAA
$\quad \neg P(x)$	(16)	$\exists$-Elimination
$\quad R(x) \to \neg P(x)$	(17)	$\to$-Introduction
$\forall x\,(R(x) \to \neg P(x))$	(18)	$\forall$-Introduction

Here is a similar example, using a function symbol this time.

Example 9.18 Let L be the first-order language with unary relation symbols R, S, and a unary function symbol f. Then starting from statements $\exists x\,R(x)$, $\forall x\,(R(x) \to S(f(x)))$ and $\forall x\,\neg(R(x) \wedge S(x))$ we may deduce $\exists x \exists y\,\neg(x = y)$.

Formal proof

$\exists x\,R(x)$	(1)	Given
$\forall x\,(R(x) \to S(f(x)))$	(2)	Given
$\forall x\,\neg(R(x) \wedge S(x))$	(3)	Given
$\quad \neg \exists x \exists y\,\neg(x = y)$	(4)	Assumption
$\quad\quad$ Let a satisfy $R(a)$	(5)	
$\quad\quad R(a) \to S(f(a))$	(6)	$\forall$-Elimination
$\quad\quad S(f(a))$	(7)	$\to$-Elimination
$\quad\quad \ldots$		

Now we use the Substitution Rule to show $f(a)$ is not equal to a.

Formal proof (continued)

$(f(a) = a)$	(8)	Assumption
$S(a)$	(9)	Substitution
$R(a) \land S(a)$	(10)	$\land$-Introduction
$\neg(R(a) \land S(a))$	(11)	$\forall$-Elimination
$\bot$	(12)	Contradiction
$\neg(f(a) = a)$	(13)	RAA
$\exists y \neg (f(a) = y)$	(14)	$\exists$-Introduction
$\exists x \exists y \neg (x = y)$	(15)	$\exists$-Introduction
$\bot$	(16)	Contradiction
$\bot$	(17)	$\exists$-Elimination
$\neg\neg \exists x \exists y \neg (x = y)$	(18)	RAA
$\exists x \exists y \neg (x = y)$	(19)	$\neg$-Elimination

The Equality Rules (Reflexivity, Symmetry, Transitivity) are more familiar as they are similar to rules you may already know about: rules concerning equivalence relations. It may be thought that equality can be regarded as really just another relation symbol, like the non-logical relation symbols, and we need only include reflexivity, transitivity and symmetry of equality. Thus one might propose that the Substitution Rule and Equality Rules in Definition 9.8 be replaced with the following statements of the first-order language

- $\forall x\,(x = x)$
- $\forall x \forall y\,(x = y \rightarrow y = x)$
- $\forall x \forall y \forall z\,((x = y \land y = z) \rightarrow x = z)$

considered as *axioms*, or additional assumptions to be used wherever needed. But the Equality and Substitution Rules do rather more than this and are quite special and powerful. We also need to know that if $x = y$ and $R(x)$ holds then $R(y)$ holds, and also that there is always only one object equal to the element x. (It would be impossible to say this using $=$ if all we knew about $=$ is that it is an equivalence relation.) The Substitution Rule is actually saying that equality is the 'finest' possible equivalence relation, and I like to think of this as saying that $=$ is a special *logical* symbol, not just a mathematical relation like an equivalence. In any case, what could be more logical than the idea of equality?

Example 9.19 The Transitivity and Symmetry Rules (two of the Equality Rules in Definition 9.8) are just special cases of the Substitution Rule.

To see this, suppose t, s are terms and we have proved $(t = s)$. Let $\theta(u, v)$ be the formula $(u = v)$. Then $(t = t)$ is provable directly using the Reflexivity Rule. Think of $(t = t)$ as the statement $\theta(t, t)$ and using the Substitution Rule substitute s for the first t in $\theta(t, t)$ using $(t = s)$ and substitute t for the second t in $\theta(t, t)$ using $(t = t)$. This gives $\theta(s, t)$ or $(s = t)$.

Similarly, suppose we have proved $(t = s)$ and $(s = r)$. Then substitute t for the t in $(t = s)$ using $(t = t)$ and substitute r for the s in $(t = s)$ using $(s = r)$. This gives $(t = r)$.

The statement $(x = x)$ is a useful one to know and convenient on many occasions because it should be true for all objects x. In other words, it behaves just like $\top$ except that it is a property of x. In the next short example it is used to show that there should always be *some* object in any interpretation of the formal system, and this example uses the Reflexivity Rule for equality in an essential way. The formal proof is a very simple two-liner.

Example 9.20 $\vdash \exists x\,(x = x)$.

Formal proof

$(x = x)$	(1)	Reflexivity
$\exists x\,(x = x)$	(2)	$\exists$-Introduction

We now move on to another very important example using the quantifier rules. This will be particularly important, especially for the proof of the Completeness Theorem later.

Example 9.21 For any formula $\theta(x)$ and any choice of distinct variables x, a there is a proof of the sentence $\exists a\, \forall x\,(\neg\,\theta(x) \vee \theta(a))$.

The difficulty here is that we would like to substitute x for a in the $\forall$-Elimination Rule, but this substitution is forbidden because a is in the scope of a $\forall x \ldots$ quantifier. In other words, the formal proof

Formal proof

$\neg \exists a\, \forall x\,(\neg\,\theta(x) \vee \theta(a))$	(1)
$\ldots$	
$\forall a\, \neg\, \forall x\,(\neg\,\theta(x) \vee \theta(a))$	(2)
$\neg\, \forall x\,(\neg\,\theta(x) \vee \theta(x))$	(3)

is disallowed for reasons similar to those in Example 9.9.

Instead, we may first change variable names and then make the required substitution.

Formal proof

$\neg\exists a\,\forall x\,(\neg\,\theta(x)\vee\theta(a))$	(1)	Assumption
$\exists b\,\forall y\,(\neg\,\theta(y)\vee\theta(b))$	(2)	Assumption
Let a satisfy $\forall y\,(\neg\,\theta(y)\vee\theta(a))$	(3)	
Let x be arbitrary	(4)	
$\neg\,\theta(x)\vee\theta(a)$	(5)	$\forall$-Elimination
$\forall x\,(\neg\,\theta(x)\vee\theta(a))$	(6)	$\forall$-Introduction
$\exists a\,\forall x\,(\neg\,\theta(x)\vee\theta(a))$	(7)	$\exists$-Introduction
$\exists a\,\forall x\,(\neg\,\theta(x)\vee\theta(a))$	(8)	$\exists$-Elimination
$\bot$	(9)	
$\neg\,\exists b\,\forall y\,(\neg\,\theta(y)\vee\theta(b))$	(10)	RAA
$\cdots$		

Note particularly the first half of the proof where we have 'unzipped' our formula using the elimination proof rules and then 'zipped' it up again with the introduction proof rules and different variables. (Other changes of variables can be achieved in the same way.)

Formal proof (continued)

$\cdots$		
Let x be arbitrary	(11)	
$\theta(x)$	(12)	Assumption
Let y be arbitrary	(13)	
$\neg\,\theta(y)\vee\theta(x)$	(14)	$\vee$-Introduction
$\forall y\,(\neg\,\theta(y)\vee\theta(x))$	(15)	$\forall$-Introduction
$\exists b\,\forall y\,(\neg\,\theta(y)\vee\theta(b))$	(16)	$\exists$-Introduction
$\bot$	(17)	
$\neg\,\theta(x)$	(18)	RAA
$\neg\,\theta(x)\vee\theta(a)$	(19)	$\vee$-Introduction
$\forall x\,(\neg\,\theta(x)\vee\theta(a))$	(20)	$\forall$-Introduction
$\exists a\,\forall x\,(\neg\,\theta(x)\vee\theta(a))$	(21)	$\exists$-Introduction
$\bot$	(22)	
$\neg\neg\,\exists a\,\forall x\,(\neg\,\theta(x)\vee\theta(a))$	(23)	RAA
$\exists a\,\forall x\,(\neg\,\theta(x)\vee\theta(a))$	(24)	

Note that in the above proof the final contradiction comes from the *original* version of the assumption, so both it and the renamed version are required.

As is often the case with formal proofs, there is rarely a single proof for a particular statement, and in this case a quite different alternative can be given.

Formal proof

$\neg \exists a \, \forall x \, (\neg \theta(x) \vee \theta(a))$	(1)	Assumption
Let a be arbitrary	(2)	
$\theta(a)$	(3)	Assumption
Let x be arbitrary	(4)	
$(\neg \theta(x) \vee \theta(a))$	(5)	$\vee$-Introduction
$\forall x \, (\neg \theta(x) \vee \theta(a))$	(6)	$\forall$-Introduction
$\exists a \, \forall x \, (\neg \theta(x) \vee \theta(a))$	(7)	$\exists$-Introduction
$\perp$	(8)	
$\neg \theta(a)$	(9)	RAA
$\forall a \, \neg \theta(a)$	(10)	$\forall$-Introduction
Let x be arbitrary	(11)	
$\neg \theta(x)$	(12)	$\forall$-Elimination
$(\neg \theta(x) \vee \theta(a))$	(13)	$\vee$-Introduction
$\forall x \, (\neg \theta(x) \vee \theta(a))$	(14)	$\forall$-Introduction
$\exists a \, \forall x \, (\neg \theta(x) \vee \theta(a))$	(15)	$\exists$-Introduction
$\perp$	(16)	
$\neg\neg \exists a \, \forall x \, (\neg \theta(x) \vee \theta(a))$	(17)	RAA
$\exists a \, \forall x \, (\neg \theta(x) \vee \theta(a))$	(18)	$\neg$-Elimination

Writing proofs is sometimes difficult and there are no prizes for the 'best' proof obtained by spotting the 'right' assumptions to make at the right time. It is therefore helpful to know all the techniques so that you can obtain some proof (any proof!) more quickly.

We are going to investigate Soundness and Completeness Theorems for this proof system. To do this, we first need to attach some sort of meaning or interpretation for a first-order language. Some of the basic ideas have been mentioned already, but it is time to be more precise. The mathematical structures we will consider include groups, fields, posets and boolean algebras, and generalise these in the sense that they have a domain of elements and various functions and relations.

Definition 9.22 Let L be a first-order language. An *L-structure* or *structure for L* is some mathematical structure $(M, \ldots, c, \ldots, f, \ldots, R, \ldots)$ where M is a non-empty set (called the *domain*), there is a distinguished named element $c \in M$ for each constant symbol c of L, there is a function $f: M^n \to M$ for each function symbol f of arity n in L (so, in particular M is closed under this function), and there is a relation $R \subseteq M^m$ for each relation symbol R of L of arity m.

As is common mathematical practice, we often talk about 'the structure M' rather than 'the structure $(M, \ldots, c, \ldots, f, \ldots, R, \ldots)$' when the operations on M are understood. Note too that our normal notation confuses the distinction between the symbol in L for a function f, constant c or relation R with the actual operation in M that interprets this symbol. Again this is standard mathematical shorthand, but would not satisfy a pedantic logician. If absolutely necessary, authors may use a different font or face, or use underlining or some other typographical device, for the symbol, as opposed to its interpretation in a structure.

Example 9.20 shows why we insist that our structures always have non-empty domain. If we allowed empty domains the Soundness Theorem (Theorem 9.25 below) would be false.

A sentence of L is one with no free variables, so must be either true or false in any L-structure. To determine which, we need to interpret the quantifiers $\forall x \ldots$ and $\exists x \ldots$ as ranging over all possible values x of the domain of M. This is possibly the most important feature of first-order logic: that all variables range over the same set of individuals. This feature distinguishes first-order logic from second-order logic (where it is possible to have variables ranging over subsets of the domain) or higher-order logics with sets of sets, and so on.

Definition 9.23 Let L be a first-order language and M an L-structure. If σ is a sentence of L, we write $M \vDash \sigma$ for 'σ is true in M'. If Σ is a set of sentences we write $M \vDash \Sigma$ to mean $M \vDash \sigma$ for all $\sigma \in \Sigma$.

If an L-formula ϕ has free variables, these variables must be given some meaning or interpretation in an L-structure M before we can say what it means for $M \vDash \phi$. In other words, these free variables should be replaced by constant symbols or closed terms with specific meaning in M. However, if some meaning is defined or understood for these variables, and we are arguing in an informal sense rather than a picky and pedantic way, we can use this understood meaning to make sense of $M \vDash \phi$ for such formulas ϕ too.

Structures are used to define the notion $\Sigma \vDash \tau$, which is essential for the statement of the Soundness and Completeness Theorems.

Definition 9.24 Let L be a first-order language, Σ a set of L-sentences and τ a further L-sentence. Then we write $\Sigma \vDash \tau$ to mean: for all L-structures M, if $M \vDash \Sigma$ then $M \vDash \tau$.

As before, $\Sigma \vDash \tau$ is vacuously true if there are no L-structures M making $M \vDash \Sigma$ true. In this case, $\Sigma \vDash \tau$ for any τ whatsoever.

Theorem 9.25 (Soundness) *Let L be a first-order language, Σ a set of L-sentences and τ a further L-sentence. Then $\Sigma \vdash \tau$ implies $\Sigma \vDash \tau$.*

Sketch-proof of Soundness Theorem Formally, the proof is by induction on the (finite) length of the derivation. We simply need to inspect each proof rule to check that it preserves the property '$\Sigma \vdash \tau$ implies $\Sigma \vDash \tau$' for all Σ, τ. □

I am not going to prove the Soundness Theorem in any more detail here. A full proof would involve a more formal definition of what it really means for an L-sentence θ to be true in an L-structure M. This can be done (by induction on the number of symbols in θ) but is unenlightening for the beginner and only becomes really useful in much more advanced work. Instead, I shall appeal to the reader's common sense and mathematical experience that the rules just given look like they should be correct and are in fact particular proof rules that he/she is accustomed to using anyway.

The Soundness Theorem is really a piece of mathematical book-keeping, checking that our proof rules are reasonable. It is chiefly used in its contrapositive form, $\Sigma \nvDash \tau$ implies $\Sigma \nvdash \tau$, which gives us a version of a very familiar technique to show something is *not* provable, to wit: find counter-example M in which Σ is true but τ is not. Then M shows that $\Sigma \nvDash \tau$, so $\Sigma \nvdash \tau$.

9.2 Examples and exercises

In exercises that use the $\to$ symbol, use the additional Rules of $\to$-Introduction and Elimination given in Remark 9.11.

Exercise 9.26 Express the following in the first-order language for the reals with non-logical symbols 0, 1, $+$, $\times$, $-$, $<$.

  (i) There is a square-root of 2.
  (ii) Every non-zero number has a multiplicative inverse.
  (iii) Some negative number has a square-root.

Exercise 9.27 Consider the following:

  (i) $\forall v \, (\theta(v) \to \psi(v))$
  (ii) $\forall v \, \theta(v) \to \forall v \, \psi(v)$
  (iii) $\exists v \, \theta(v) \to \exists v \, \psi(v)$

Prove that (i) implies (ii). Prove as many other implications between these three statements as you can.

Exercise 9.28 Let L be the first-order language with a unary function symbol f and constants 0, 1. Prove that from the sentences

$$\forall x\,(f(x) = 0 \vee f(x) = 1)$$

and

$$\exists x\,\exists y\,\exists z\,(\neg x = y \wedge \neg y = z \wedge \neg z = x)$$

you may deduce

$$\exists x\,\exists y\,(\neg x = y \wedge f(x) - f(y)).$$

Exercise 9.29 Let L be the first-order language with a unary function symbol f. Prove that from the sentences

$$\forall x\,(f(f(x)) = x),$$

$$\forall x\,\forall y\,\forall z\,\forall w\,(x = y \vee x = z \vee x = w \vee y = z \vee y = w \vee z = w),$$

and

$$\exists x\,\exists y\,\exists z\,(\neg x = y \wedge \neg y = z \wedge \neg z = x)$$

you may deduce

$$\exists u\, f(u) = u.$$

Exercise 9.30 A (simple) *graph* is usually defined as a non-empty set V of vertices together with a set E of edges, which are considered as unordered pairs of vertices. Define a graph as a first-order structure where the domain is the set of vertices and the edges are represented by a binary relation symbol E, and $E(x, y)$ holds if and only if there is an edge from x to y.

Which first-order axiom(s) is(are) also required for your graphs to correspond to the usual ones?

Exercise 9.31 Write down statements in the first-order language of graphs of Exercise 9.30 stating the following.

(i) Between and two vertices there is a path of length at most 4.
(ii) There is a *cycle* of length 4 but no cycle of length 3.

(iii) The graph has a *clique* (complete subgraph) of 5 or more vertices.

(iv) The graph is not complete (i.e. not every pair of vertices is connected by an edge.)

Exercise 9.32 Write down the following axioms in the first-order languages indicated. State the arities of each function and relation symbol. If you use any abbreviations, say what they are.

(i) The theory of groups (constant for e; binary function for $\times$; unary function for $^{-1}$).

(ii) The theory of fields (constants 0, 1; binary functions $+$, $\times$).

(iii) The theory of an equivalence relation that has exactly five equivalence classes (no constants or functions; binary relation symbol R for the equivalence relation).

(iv) The theory of linear orders with no end-points (binary relation $\leqslant$ for the order relation).

(v) The theory of a poset with an identified chain as a subset (binary relation $\leqslant$ for the order; unary relation C for the chain).

Exercise 9.33 (This exercise is a long list of equivalences you might like to refer back to later. You do not necessarily have to do them all, just a handful to check you understand what is going on. We say that the *dual* of $\forall$ is $\exists$, and the *dual* of $\exists$ is $\forall$.)

Prove that the following are logically equivalent for all ϕ, ψ and all variables x as indicated. You can consider each equivalence syntactically in the formal proof system, or semantically by considering structures directly. In cases with provisos about variables not being free, give an example to show why the condition added is necessary.

- $\neg \exists x \, \phi$, $\forall x \neg \phi$.
- $\neg \forall x \, \phi$, $\exists x \neg \phi$.
- $\neg \neg \phi$, ϕ.
- $\forall x \, (\phi \wedge \psi)$, $(\forall x \, \phi \wedge \psi)$, provided x is not free in ψ.
- $\forall x \, (\phi \wedge \psi)$, $(\phi \wedge \forall x \, \psi)$, provided x is not free in ϕ.
- $\exists x \, (\phi \wedge \psi)$, $(\exists x \, \phi \wedge \psi)$, provided x is not free in ψ.
- $\exists x \, (\phi \wedge \psi)$, $(\phi \wedge \exists x \, \psi)$, provided x is not free in ϕ.
- $Qx(\phi \vee \psi)$, $(Qx\phi \vee \psi)$, provided x is not free in ψ, $Q = \forall$ or $\exists$.
- $Qx(\phi \vee \psi)$, $(\phi \vee Qx\psi)$, provided x is not free in ϕ, $Q = \forall$ or $\exists$.
- $Qx(\phi \rightarrow \psi)$, $(\phi \rightarrow Qx\psi)$, provided x is not free in ϕ, $Q = \forall$ or $\exists$.
- $Qx(\phi \rightarrow \psi)$, $(Q^*x\phi \rightarrow \psi)$, provided x is not free in ψ, $Q = \forall$ or $\exists$, and Q^* is the dual of Q.

Exercise 9.34 Use the previous exercise and induction to show that every formula $\psi(x_1, \ldots, x_k)$ in some first-order language L is equivalent to a formula in *prenex normal form*, i.e. to a formula in the same language L of the form $Q_1 y_1 \ldots Q_l y_l \theta(x_1, \ldots, x_k, y_1 \ldots, y_l)$ where each Q_i is $\forall$ or $\exists$ and the formula $\theta(x_1, \ldots, x_k, y_1 \ldots, y_l)$ contains no quantifiers.

Exercise 9.35 (a) Show that from $\forall x \exists y \forall z ((\neg R(x, y) \wedge S(y, z)) \vee R(x, z))$ you may prove $\forall v \exists w \forall x ((\neg R(v, w) \wedge S(w, x)) \vee R(v, x))$.

(b) Explain in as general terms as you can how the proof rules for first-order logic enable you to rename all bound variables in a sentence σ (providing there are no name clashes or illegal substitutions, of course).

(c*) Attempt to state and prove a general theorem on renaming variables in formulas. (The proof will be by induction on the complexity of formulas.)

Exercise 9.36 Give a modified set of proof rules for first-order logic that allow for empty domains. Check that the Soundness Theorem holds for your logic and that, on adding the single statement $\exists x (x = x)$ as an axiom, all the provable statements are exactly those of the normal first-order logic.

9.3 Second- and higher-order logic*

Calling our logic 'first-order' begs the question as to what 'second-order' logic is.

The essential feature of first-order logic is that the quantifiers $\forall x \ldots$ and $\exists x \ldots$ are only allowed to range over *elements* of a non-empty domain, the domain of the structure M under consideration. This limits the things one can say quite considerably. For example it is possible to say that a structure G is a group in first-order logic, but it is not possible to say that G is a simple group. The obvious attempts fail because the usual definition is that G is a group if and only if it has no *subset* $N \subseteq G$ containing the identity and closed under certain operations (multiplication, inverse, and conjugation by arbitrary elements of G) other than $\{1\}$ or G itself. This seems to require a quantifier over subsets of the structure, and first-order logic does not have this.

The reader should be very wary of any false sense of security in these arguments. They are sometimes rather deep and often very difficult. I did not prove anything in the last paragraph – only pointed out that the obvious attempt to say in first-order logic that 'G is simple' fails. At this stage, it seems possible that there are other less obvious approaches. In fact, it really cannot be done, but we will not have the apparatus to prove this until the next chapter. (See

Exercise 10.17 for a hint.) In contrast, some cases can be done: we can describe all simple groups of order 168 (or any other finite order) by a first-order sentence, for example.

Second-order logic allows for these subset-quantifiers. A particular language for second-order logic is obtained from one for first-order logic by adding a new set of variables X_i^j for each $i, j \in \mathbb{N}$. The idea is that X_i^j ranges over subsets of M^j, where M is the structure under consideration. We also add new atomic formulas $(t_1, \ldots, t_j) \in X_i^j$ for every i, j and terms $t_1, \ldots, t_j$. We allow these new atomic formulas to take part in more complex formulas and build up the notion of a second-order formula in a similar way to that for first-order formulas. The structures for a second-order language are the same as those for the corresponding first-order language.

In fact, although we have added second-order variables X_i^j of all 'types' j, thinking of X_i^j as ranging over subsets of M^j, we never need more than the first three cases, $j = 1, 2, 3$. This is because there is always a bijection from an infinite set M to its cartesian power M^2.

Exercise 9.37 Assume the assertion just made, that for any infinite set M there is a bijection $f: M \to M^2$. Write down a formula in second-order logic that expresses the fact that X_1^3 is the set $\{(x, y, z) : z = f(x, y)\}$ for some such function f. Use this to show that second-order variables of 'type' greater than three are unnecessary.

This then gives the main ideas of second-order logic. We have a language, a grammar explaining how formulas are built, and a notion of what it means for a second-order sentence to be true in a structure M. (For the latter all we need to know is that $M \vDash \exists X_i^j \, \theta(X_i^j)$ holds if and only if $M \vDash \theta(A)$ for some $A \subseteq M^j$ and $M \vDash \forall X_i^j \, \theta(X_i^j)$ holds if and only if $M \vDash \theta(A)$ for all $A \subseteq M^j$, and use the other ideas from first-order logic.)

If this seems interesting, then good: it *is* interesting – in fact, far *too* interesting! Some structures, such as those for the reals and the naturals, can be characterised completely by the second-order sentences true in them, and sentences achieving this were first written down by Dedekind and, in the case of the naturals, independently by Peano. The real problem comes when one tries to put right the most conspicuous omission in the account of second-order logic: the lack of any system for proof. There is no such system, and in fact it is a consequence of Gödel's famous theorem on the incompleteness of arithmetic that there *cannot* be such a second-order system for $\mathbb{N}$ that is any better than simply writing down all the true sentences and using them as a set of axioms. (This is not what we want: we want a way of discovering *new* true statements!)

There are possible ways round at least some of the problems here. One is to change our idea of structure to allow two or more domains: one of numbers or objects, and another of sets of objects, for example. Thus the structure for the natural numbers becomes $(\mathbb{N}, P(\mathbb{N}), P(\mathbb{N}^2), P(\mathbb{N}^3), \ldots, 0, 1, \ldots)$, where $P(S)$ is the power-set or set of all subsets of S. We can even write down some suitable axioms, such as the axiom of extensionality saying two sets are equal if and only if their elements (in the sense of the symbol $\in$) are the same, and also lists of axioms – called *axiom schemes* – expressing the existence of plenty of sets, such as

$$\exists X_1^1 \, \forall x \, (x \in X_1^1 \leftrightarrow \theta(x))$$

for all second-order formulas $\theta(x)$. (This particular axiom scheme is called comprehension.)

As a tool for proving statements about the naturals, the approach just suggested is quite powerful. But in fact, it is just first-order logic in disguise once again. Consider structures of the form

$$(\mathbb{N} \cup P(\mathbb{N}) \cup P(\mathbb{N} \times \mathbb{N}) \cup \ldots, N, S_1, S_2, S_3, \ldots, \in_1, \in_2, \in_3, \ldots, 0, 1, \ldots)$$

where N, S_1, S_2, S_3 are all unary relations: $N(x)$ means 'x is a number' or $x \in \mathbb{N}$; $S_1(x)$ means 'x is a set' or $x \subseteq \mathbb{N}$; $S_2(x)$ means 'x is a set of pairs' or $x \subseteq \mathbb{N} \times \mathbb{N}$; and $S_3(x)$ means 'x is a set of triples'; etc. The $k+1$-ary relation $(x_1, \ldots, x_k) \in_k s$ means 'the k-tuple $(x_1, \ldots, x_k)$ is a member of the set s.'.

A similar approach can be used for even stronger third-order logic where variables ranging over sets of sets are introduced, or for higher-order logic. An even more powerful approach is to adopt a first-order set theory such as Zermelo–Fränkel as a base theory to prove statements about $\mathbb{N}$. This behaves rather like nth order logic for all n, including transfinite n, and amazingly is a first-order theory in the language with a single binary relation symbol, $\in$. Thus we return, very squarely, back to the realm of first-order logic.

10

Completeness and compactness

10.1 Proof of completeness and compactness

This chapter is devoted to the main theorem for first-order logic, the Completeness Theorem.

Theorem 10.1 (Completeness) *Let L be a first-order language, Σ a set of L-sentences and τ a further L-sentence. Then $\Sigma \vDash \tau$ implies $\Sigma \vdash \tau$.*

By contrast with the Soundness Theorem, the Completeness Theorem for first-order logic is a powerful mathematical tool with very many interesting consequences and applications. Its proof involves an application of Zorn's Lemma in an essential way, and will take us quite a bit longer. We start by looking at a way of adding new constant symbols to the language, called 'Henkinisation', and named after the logician Leon Henkin who first used this method.

Definition 10.2 Let L be a first-order language. We consider for each formula $\phi(x)$ with a single free variable x a new constant symbol ε. ('New' here means that these constants are distinct symbols not already in L. We only need a supply of new symbols and we may assume that there are always enough new symbols available.) Since there will be one such constant ε for each $\phi(x)$ and we require these symbols to be all distinct we shall denote the symbol corresponding to $\phi(x)$ by $\varepsilon_{\phi(x)}$. This is just a name for a new symbol which is used to indicate this symbol's use in the formulas below. There is no new logical rule being introduced here.

The *first Henkinisation of L* is the first-order language $H(L)$ consisting of L together with all $\varepsilon_{\phi(x)}$.

The *second Henkinisation of L* is the first-order language $H(H(L))$ consisting of $H(L)$ together with all $\varepsilon_{\phi(x)}$ for formulas ϕ and variables x in the first Henkinisation.

The *complete Henkinisation of L* is the first-order language L_H consisting of all symbols in $L, H(L), H(H(L)), \ldots$.

The *Henkin axiom* corresponding to a Henkin constant $\varepsilon_{\phi(x)}$ is the first-order statement $\forall x (\neg \phi(x) \vee \phi(\varepsilon_{\phi(x)}))$ of the appropriate Henkinised language.

The idea is that we add new constants $\varepsilon_{\phi(x)}$ with the property (expressed by the Henkin axiom) that says if any element satisfies $\phi(x)$ then $\varepsilon_{\phi(x)}$ does.

Lemma 10.3 *Let L be a first-order language, and suppose Σ is a set of first-order L-sentences which is consistent, i.e. $\Sigma \nvdash \bot$. Let $H(L)$ be the first Henkinisation of L and H_1 the set of Henkin axioms for the Henkin constants added. Then $\Sigma \cup H_1$ is also consistent.*

Proof We suppose otherwise, that $\Sigma \cup H_1 \vdash \bot$, and show this implies that $\Sigma \vdash \bot$. By assumption there is a finite proof of $\bot$ from $\Sigma \cup H_1$, so for finitely many formulas ϕ_i of L and finitely many variables x_i of L we have

$$\Sigma \cup \left\{ \forall x_1 (\neg \phi_1(x_1) \vee \phi_1(\varepsilon_{\phi_1(x_1)})), \ldots, \forall x_k (\neg \phi_k(x_k) \vee \phi_k(\varepsilon_{\phi_k(x_k)})) \right\} \vdash \bot.$$

The Henkin constants do not appear anywhere except where indicated, and behave in the proof just like variables, so the $\exists$-Elimination rule applies. By k applications of this rule we have

$$\Sigma \cup \left\{ \exists y_1 \forall x_1 (\neg \phi_1(x_1) \vee \phi_1(y_1)), \ldots, \exists y_k \forall x_k (\neg \phi_k(x_k) \vee \phi_k(y_k)) \right\} \vdash \bot.$$

But by Example 9.21, we have that

$$\vdash \exists y_i \forall x_i (\neg \phi_i(x_i) \vee \phi_i(y_i))$$

for each i so therefore $\Sigma \vdash \bot$, as required. $\qquad\square$

Lemma 10.4 *Let L be a first-order language, and suppose Σ is a set of first-order L-sentences which is consistent, i.e. $\Sigma \nvdash \bot$. Let L_H be the complete Henkinisation of L and H the set of Henkin axioms for all new Henkin constants. Then $\Sigma \cup H$ is also consistent.*

Proof If not, $\Sigma \cup H \vdash \bot$, so there is a finite proof of $\bot$ from $\Sigma \cup H$ and this proof must use finitely many Henkin constants so can be expressed entirely in the nth Henkinisation $H(\cdots(H(L))\cdots)$ of L, for some n. But this is impossible by n applications of the previous lemma. $\qquad\square$

Lemma 10.5 *Let L be a first-order language, and suppose* Σ *is a consistent set of first-order L-sentences, and* θ *is a further L-sentence. Then at least one of* $\Sigma \cup \{\theta\}$, $\Sigma \cup \{\neg\theta\}$ *is also consistent.*

Proof If $\Sigma \cup \{\theta\} \vdash \perp$ then by Reductio Ad Absurdum $\Sigma \vdash \neg\theta$. But then we cannot have $\Sigma \cup \{\neg\theta\} \vdash \perp$, for this would mean that $\Sigma \vdash \perp$. So either $\Sigma \cup \{\theta\} \nvdash \perp$ or $\Sigma \cup \{\neg\theta\} \nvdash \perp$. $\square$

We are now in a position to prove the Completeness Theorem.

We shall suppose that $\Sigma \nvdash \tau$ and find a structure M making Σ true but not τ. It will follow that $\Sigma \nvDash \tau$, thus proving the contrapositive of the Completeness Theorem.

By a now familiar application of the Reductio Ad Absurdum Rule, $\Sigma \nvdash \tau$ implies $\Sigma \cup \{\neg\tau\}$ is consistent. If we can show that there is a structure M making $\Sigma \cup \{\neg\tau\}$ true it would follow that $M \vDash \Sigma$ and $M \nvDash \tau$ hence $\Sigma \nvDash \tau$, as required.

Let L_H be the complete Henkinisation of L, and H the set of Henkin axioms. By the lemma on Henkinisation, $\Sigma_0 = \Sigma \cup H \cup \{\neg\tau\}$ is consistent. We look at the set X of all consistent sets of L_H sentences $\Xi \supseteq \Sigma_0$, and order X by $\subseteq$. This makes X into a poset with the Zorn property since any chain $Y \subseteq X$ has upper bound $\Xi = \bigcup Y = \{\sigma : \sigma \in \Gamma, \text{ for some } \Gamma \in Y\}$. (The argument that this is consistent has been given before and simply requires that proofs are finite objects and Y is a chain.) Therefore there is a set $\Sigma^+ \in X$ which is consistent, contains Σ_0, and is maximal. By maximality and the previous lemma, $\Sigma^+ \in X$ has the property that for all sentences θ of L_H we have either $\theta \in \Sigma^+$ or $\neg\theta \in \Sigma^+$.

The set Σ^+ contains all the information we need. Notice that if σ is any L_H-sentence and $\Sigma^+ \vdash \sigma$ then $\sigma \in \Sigma^+$. The process from now on requires defining our L-structure M and checking various properties such as well-definedness, and that the structure we define satisfies the sentences in Σ_0. Each step in the process uses the maximality of Σ^+ and will also require applications of the individual proof rules for our system.

First, we need to define M. The domain of M is defined to be the set T of all closed terms of L_H, factored out by the equivalence relation

$$t \sim s \text{ if and only if } t = s \in \Sigma^+.$$

Clearly the first task is to show this is an equivalence relation. This uses the equality rules of first-order logic. If $t \in T$ then $\vdash t = t$, so $t = t \in \Sigma^+$ hence $t \sim t$ and $\sim$ is reflexive. If $t, s \in T$ and $t \sim s$ then $t = s \in \Sigma^+$ so $\Sigma^+ \vdash s = t$ and hence

$s \sim t$. Also if $t, s, r \in T$ and $t \sim s$ and $s \sim r$ then $t = s, s = r \in \Sigma^+$ so $\Sigma^+ \vdash t = r$ hence $t \sim r$. Thus $\sim$ is an equivalence.

Write $[t]$ for the equivalence class of $t \in T$. We must next define the constants, relations and functions on M. This is done as follows:

- for each constant symbol c, the corresponding element of M is $[c]$;
- for each n-ary relation symbol R, we define the meaning of R on M by $R([t_1], \ldots, [t_n])$ if and only if $R(t_1, \ldots, t_n) \in \Sigma^+$;
- for each n-ary function symbol f, we define the meaning of f in M by $f([t_1], \ldots, [t_n]) = [f(t_1, \ldots, t_n)]$.

Our next task is to show that the functions and relations here are well defined and do not depend on the choice of representatives $t_1, \ldots, t_n$ of $[t_1], \ldots, [t_n]$. There is nothing to do for the constant symbols. For a relation symbol R we must prove that if $R(t_1, \ldots, t_n) \in \Sigma^+$ and $s_i \sim t_i$ for each i then $R(s_1, \ldots, s_n) \in \Sigma^+$. But this follows from maximality: $R(t_1, \ldots, t_n) \in \Sigma^+$ and $s_i = t_i \in \Sigma^+$ for each i implies that $\Sigma^+ \vdash R(s_1, \ldots, s_n)$ by the Substitution Rule. The treatment of function symbols is similar; here we must prove that if $s_i \sim t_i$ for each i then $f(t_1, \ldots, t_n) \sim f(s_1, \ldots, s_n)$. Applying the Substitution Rule to the formula $\theta(x_1, \ldots, x_n)$ defined to be $f(t_1, \ldots, t_n) = f(x_1, \ldots, x_n)$ we have $\Sigma^+ \vdash \theta(t_1, \ldots, t_n)$, by the Reflexivity Rule, hence $\Sigma^+ \vdash f(t_1, \ldots, t_n) = f(s_1, \ldots, s_n)$ by substitution.

Finally we have to show that M makes Σ_0 true. In fact, as in the proof of the Completeness Theorem for propositional logic, it is easier to show that M makes Σ^+ true, and we do this by an induction on the 'size' of a sentence θ. For the purposes of this proof, we say a *connective* is one of the symbols $\neg, \vee, \wedge, \forall, \exists$. Then we show inductively that

- for all θ with at most n *connectives* we have: $M \vDash \theta$ if and only if $\theta \in \Sigma^+$.

If θ has 0 connectives, it may be $t = s$ for some $t, s \in T$. Observe that it is quite possible for two different closed terms $t, s \in T$ to represent the same object in M. But by the definition of M, $M \vDash t = s$ holds if and only if $t \sim s$ which, by the definition of $\sim$, holds if and only if $t = s \in \Sigma^+$.

The other possibility when θ has 0 connectives is when θ is $R(t_1, \ldots, t_n)$, for some relation symbol R and some terms t_i. But then, by the definition of R in M, $M \vDash R(t_1, \ldots, t_n)$ if and only if $R(t_1, \ldots, t_n) \in \Sigma^+$.

We now suppose θ has one or more connectives and use our induction hypothesis on formulas with fewer connectives.

If θ is $\neg\phi$ then using the induction hypothesis the following are equivalent: $M\vDash\neg\phi$; $M\nvDash\phi$; $\phi\notin\Sigma^+$; and $\neg\phi\in\Sigma^+$. The last step here is by the maximality of Σ^+.

If θ is $\phi\wedge\psi$ then $M\vDash\theta$ if and only if $M\vDash\phi$ and $M\vDash\psi$, which by the induction hypothesis is true if and only if $\phi\in\Sigma^+$ and $\psi\in\Sigma^+$. But then, by an application of $\wedge$-Introduction, $\phi\in\Sigma^+$ and $\psi\in\Sigma^+$ imply $\phi\wedge\psi\in\Sigma^+$; the converse is similar using $\wedge$-Elimination just as in the proof of the Completeness Theorem for propositional logic.

If θ is $\phi\vee\psi$ we have $M\vDash\theta$ if and only if $M\vDash\phi$ or $M\vDash\psi$, which, using the induction hypothesis, is true if and only if $\phi\in\Sigma^+$ or $\psi\in\Sigma^+$. But then by a similar argument to that in the last paragraph using maximality and $\vee$-Introduction, $\phi\in\Sigma^+$ or $\psi\in\Sigma^+$ implies $\phi\vee\psi\in\Sigma^+$, and for the converse $\phi\vee\psi\in\Sigma^+$ implies $\phi\in\Sigma^+$ or $\psi\in\Sigma^+$ since if $\phi\notin\Sigma^+$ and $\psi\notin\Sigma^+$ then $\neg\phi\in\Sigma^+$ and $\neg\psi\in\Sigma^+$ so $\Sigma^+,\phi\vee\psi\vdash\bot$ by the $\vee$-Elimination Rule and the Contradiction Rule.

Since the domain of M is the set of equivalence classes of closed terms from T and each term represents its own equivalence class we have $M\vDash\forall x\,\phi(x)$ if and only if $M\vDash\phi(t)$ for each $t\in T$. Now if $\forall x\,\phi(x)\in\Sigma^+$ and $\phi(x)$ has at most n connectives, then $\Sigma^+\vdash\phi(t)$ for all t by $\forall$-Elimination, so $M\vDash\forall x\,\phi(x)$ by the maximality of Σ^+ and the induction hypothesis. For the converse, we use the Henkin axiom for $\neg\phi(x)$. Suppose $\forall x\,\phi(x)\notin\Sigma^+$, so $\neg\forall x\,\phi(x)\in\Sigma^+$ by maximality, and recall that $\forall x\,(\neg\neg\phi(x)\vee\neg\phi(\varepsilon_{\neg\phi(x)}))$ is in Σ^+. Then the following is a valid proof from Σ^+.

Formal proof

$\neg\forall x\,\phi(x)$	(1)	Given, in Σ^+
$\quad\neg\neg\phi(\varepsilon_{\neg\phi(x)})$	(2)	Assumption
$\quad$ Let x be arbitrary	(3)	
$\quad\quad\neg\phi(x)$	(4)	Assumption
$\quad\quad\neg\neg\phi(x)\vee\neg\phi(\varepsilon_{\neg\phi(x)})$	(5)	$\forall$-Elimination
$\quad\quad\neg\neg\phi(x)$	(6)	$\vee$-Elimination
$\quad\quad\bot$	(7)	
$\quad\quad\neg\neg\phi(x)$	(8)	RAA
$\quad\quad\phi(x)$	(9)	
$\quad\forall x\,\phi(x)$	(10)	$\forall$-Introduction
$\quad\bot$	(11)	
$\neg\neg\neg\phi(\varepsilon_{\neg\phi(x)})$	(12)	RAA
$\neg\phi(\varepsilon_{\neg\phi(x)})$	(13)	

This shows that $\varepsilon_{\neg\phi(x)}$ is a $t \in T$ such that $\phi(t) \notin \Sigma^+$, so $M \nvDash \phi(t)$ by the induction hypothesis. Hence $M \nvDash \forall x\,\phi(x)$, as required.

The case of the $\exists$ quantifier is similar. For one direction, if $M \vDash \exists x\,\phi(x)$ where $\phi(x)$ has at most n connectives then $M \vDash \phi(t)$ for some $t \in T$, so $\phi(t) \in \Sigma^+$ by the induction hypothesis, and hence $\exists x\,\phi(x) \in \Sigma^+$ by maximality of Σ^+ and $\exists$-Introduction. Conversely, if $\exists x\,\phi(x) \in \Sigma^+$ then since we also have $\forall x\,(\neg\phi(x) \vee \phi(\varepsilon_{\phi(x)})) \in \Sigma^+$ the following is a proof from Σ^+.

Formal proof

$\exists x\,\phi(x)$	(1)	Given, in Σ^+
Let a satisfy $\phi(a)$	(2)	
$\neg\phi(a) \vee \phi(\varepsilon_{\phi(x)})$	(3)	$\forall$-Elimination
$\quad \neg\phi(a)$	(4)	Assumption
$\quad \bot$	(5)	
$\neg\neg\phi(a)$	(6)	RAA
$\phi(\varepsilon_{\phi(x)})$	(7)	$\vee$-Elimination
$\phi(\varepsilon_{\phi(x)})$	(8)	$\exists$-Elimination

So the term $t = \varepsilon_{\phi(x)}$ has $\phi(t)$ in Σ^+ hence $M \vDash \phi(t)$ by induction, and so $M \vDash \exists x\,\phi(x)$, as required.

This completes the inductive proof that $M \vDash \Sigma^+$ and hence completes the proof of the Completeness Theorem.

The Completeness and Soundness Theorems are interesting from a foundational point of view (where they show that just about any mathematical proof can be expressed in a formal system which is mechanically checkable), but are also powerful mathematical tools in their own right. We will be investigating them as mathematics in the next two chapters, but we can give here a sample corollary, the Compactness Theorem. It is a powerful result allowing us to construct new mathematical structures, and is proved by using the Soundness and Completeness Theorems together to allow us to pass between the worlds of proofs and mathematical structures.

Theorem 10.6 (Compactness) *Let L be a first-order language and Σ a set of L-sentences. Suppose that each finite subset Γ of Σ has a structure $M_\Gamma \vDash \Gamma$ making Γ true. Then there is a structure M making the whole of Σ true.*

Proof By assumption and the Soundness Theorem, each finite $\Gamma \subseteq \Sigma$ is consistent, $\Gamma \nvdash \bot$. This means Σ is consistent too, since a proof of $\bot$ from Σ is finite so can only use finitely many assumptions from Σ. So $\Sigma \nvdash \bot$. It follows from the Completeness Theorem that there is a structure M making the whole of Σ true.   □

10.2 Examples and exercises

For the next few paragraphs and exercises, we shall examine the 'pure' first-order language L_0 with $=$ and all the other logical symbols, but no non-logical symbols.

Exercise 10.7 Indicate how to write down sentences σ_n and τ_n of L_0 with the following meanings.

- σ_n: There are at least n elements in the domain.
- τ_n: There are exactly n elements in the domain.

(Hint: use plenty of variables. Look at Exercises 9.28 and 9.29 for ideas.)

Exercise 10.8 Write out a formal proof showing that $\sigma_3 \vdash \sigma_2$.

Exercise 10.9 Explain why a structure M satisfies

$$S_\infty = \{\sigma_n : n \in \mathbb{N}\}$$

if and only if its domain is infinite.

The last exercise shows that the class of infinite M is axiomatisable in L_0, i.e. there is a set of sentences of L_0 such that M is infinite if and only if it satisfies all these sentences.

Example 10.10 The class of infinite M is not finitely axiomatisable in the language L_0, i.e. there is no single sentence σ_∞ of L_0 which is satisfied in a structure M if and only if the domain of M is infinite.

Proof If σ_∞ is such a sentence it must be that $\sigma_\infty \vdash \sigma_k$ for all k. This is by the Completeness Theorem, because clearly $\sigma_\infty \vDash \sigma_k$ for all k. Similarly, by completeness, there is a proof p of σ_∞ from the set of sentences S_∞, since once again $S_\infty \vDash \sigma_\infty$. But proofs are finite objects, so there is a finite subset of S_∞ such that

$$\sigma_1, \ldots, \sigma_n \vdash \sigma_\infty \vdash \sigma_k$$

for all k. In particular this would imply that

$$\sigma_1, \ldots, \sigma_n \vdash \sigma_{n+1}.$$

But this is impossible since there is an L_0 structure with exactly n elements, and this satisfies $\sigma_1, \ldots, \sigma_n$ but not σ_{n+1}, so

$$\sigma_1, \ldots, \sigma_n \nvDash \sigma_{n+1}$$

hence

$$\sigma_1, \ldots, \sigma_n \nvdash \sigma_{n+1}$$

by the Soundness Theorem.   □

Exercise 10.11 Show that there is no set S_f of L_0-sentences such that a structure satisfies S_f if and only if its domain is finite. (Hint: if such a set S_f exists, show that $S_f \cup \{\sigma_1, \ldots, \sigma_n\}$ is consistent for all n. Using compactness or otherwise, derive a contradiction.)

Exercise 10.12 Find a single sentence θ in a (necessarily different) first-order language L that only has infinite models.

Exercise 10.13 Show that the following properties of a graph (represented in a first-order language as in Exercise 9.30) are not equivalent to any set of first-order statements in this language.

(i) The graph is connected, i.e. between any two vertices there is a finite path connecting them.

(ii) The graph is not a forest, i.e. it has a finite cycle.

(iii) There is no infinite clique in the graph.

Example 10.14 The first-order theory of groups can be described in the language L_G with one constant symbol e for the identity, one binary function symbol $\times$ for the group multiplication and one unary function $^{-1}$ for inverses. The *theory of groups* is the set of L_G sentences

- $\forall x \forall y \forall z ((x \times y) \times z = x \times (y \times z))$
- $\forall x (x \times e = x \wedge e \times x = x)$
- $\forall x (x \times x^{-1} = e \wedge x^{-1} \times x = e)$

so that a *group* is an L_G structure satisfying these sentences. (Note that 'closure under multiplication and inverses' is automatic from our definition of L_G-structure.)

Example 10.15 A *cyclic group* is a group G generated by a single element x, that is $G = \{x^n : n \in \mathbb{Z}\}$ where $x^0 = e$ and $x^{-n} = (x^{-1})^n$ for negative exponents.

Then, in contrast with the last example, the set of cyclic groups is *not* axiomatisable, i.e. there is no set of first-order sentences true for precisely the cyclic groups.

Proof Define a set of first-order sentences involving two new constant symbols a, b and the other L_G operations by

$$\Gamma = \left\{ \neg a^n = b^k : n, k \in \mathbb{Z}, \text{not both zero} \right\}.$$

Here, a^n abbreviates some term in the language involving a, $\times$ and possibly also e (if $n = 0$) and $^{-1}$ (if $n < 0$). Then if G is a group and $a, b \in G \vDash \Gamma$ then G is not cyclic, for if G is generated by $x \in G$ we have $a = x^r$ and $b = x^s$ for some $r, s \in \mathbb{Z}$ and so $a^s = x^{rs} = b^r$ so $r = s = 0$ by Γ. But this would mean $a = b = e$ so $a^n = b^k$ for all n, k, contradicting Γ.

Now suppose Σ is a set of sentences containing the axioms for group theory and such that each $\sigma \in \Sigma$ is true in all cyclic groups. To show that Σ does not axiomatise all cyclic groups it suffices to find a non-cyclic group satisfying Σ. To this end, let $N_0 \in \mathbb{N}$ and let

$$\Gamma_0 = \left\{ \neg a^n = b^k : n, k \in \mathbb{Z}, \text{not both zero}, |n|, |k| < N_0 \right\}.$$

Now consider

$$C = \{ \ldots, x^{-2}, x^{-1}, x^0, x^1, x^2, \ldots \},$$

the infinite cyclic group generated by x. Then $C \vDash \Sigma$ as C is cyclic. Let $a = x$ and $b = x^{N_0}$. Then if $a^n = b^k$ with $|n|, |k| < N_0$ we have $kN_0 - n = 0$ which has $n = k = 0$ as its only solution. Thus $C \vDash \Gamma_0$.

The previous paragraph shows that for each finite subset $\Gamma_0 \subseteq \Gamma$ there is a group C with $a, b \in C$ such that $(C, a, b) \vDash \Gamma_0 \cup \Sigma$. By the Compactness Theorem it follows that there is a group G with elements $a, b \in G$ and $(G, a, b) \vDash \Gamma \cup \Sigma$. So this group satisfies Σ but is not cyclic as it also satisfies Γ, as required. □

The technique of going to an expanded language by adding constants and then applying the Compactness Theorem is typical, and we will see other examples like this.

Exercise 10.16 Which of the following classes of groups is axiomatisable in the language L_G of Example 10.14? Finitely axiomatisable in L_G? Give proofs.

   (i) The class of groups of size n (for some fixed $n \in \mathbb{N}$).
   (ii) The class of all groups of size at most n (for some fixed $n \in \mathbb{N}$).
   (iii) The class of all infinite groups.

(iv) The class of all finite groups.

(v) The class of all torsion groups.

(vi) The class of all n-torsion groups.

(vii) The class of all torsion-free groups.

(A group G is *n-torsion* if $g^n = e$ for all $g \in G$. It is *torsion* if for all $g \in G$ there is $n \in \mathbb{N}$ such that $g^n = e$, and it is *torsion-free* if for all $g \in G$, if $g \neq e$ then $g^n \neq e$ for all $n \in \mathbb{N}$.)

There are very many ways of doing the next exercise, and it is instructive to try several and see what extra information each gives. Possibly the simplest is to note that an abelian group is simple if and only if it is cyclic of prime order. By compactness every set of sentences true in all simple abelian groups is also true in some infinite group.

Exercise 10.17 Prove that the class of all simple groups is not axiomatisable.

10.3 The Compactness Theorem and topology*

The Compactness Theorem is not idly named. It is actually equivalent to the assertion that a particular topological space is compact. This short optional section defines the topological space and shows the connection, and is provided here for readers with some background knowledge in general topology.

First, we fix a first-order language L. We are interested in L-structures and the L-sentences that they satisfy.

Definition 10.18 Let M be an L-structure. The *theory of M*, written $\mathrm{Th}\,M$, is the set $\{\sigma : M \vDash \sigma\}$ of all L-sentences true in M.

We can see quickly that the set $\Sigma = \mathrm{Th}\,M$ has the property

- if σ is any L-sentence then exactly one of $\sigma \in \Sigma$ or $\neg\sigma \in \Sigma$.

This is just a restatement of the fact that every σ is either true or false in a structure for L, but not both.

In fact, by the Soundness Theorem, we can also see that $\mathrm{Th}\,M$ is consistent, i.e. $\mathrm{Th}\,M \nvdash \bot$, and the above property shows that it is maximally consistent, i.e. that there is no proper extension which is a set of L-sentences and consistent. Also, by the Completeness Theorem, any such maximally consistent set of L-sentences is $\mathrm{Th}\,M$ for some L-structure M. Thus the idea of the theory of M characterises such maximally consistent sets, though we will not require these observations here.

We are going to study the set of all sets Σ of L-sentences of the form $\text{Th}\,M$ where M is an L-structure. We define

$$X = \{\text{Th}\,M : M \text{ is an } L\text{-structure}\}.$$

We will define a topology on X by specifying a suitable collection of open sets. For each L-sentence σ we define

$$U_\sigma = \{\Sigma \in X : \sigma \in \Sigma\}.$$

Note that $U_\perp = \varnothing$ since no L-structure makes $\perp$ true. Similarly, $U_\top = X$ as $\top \in \text{Th}\,M$ for all M. It is more interesting to note that, for each σ,

$$X = U_\sigma \cup U_{\neg\sigma}$$

since any $\Sigma = \text{Th}\,M$ must contain either σ or $\neg\sigma$, and furthermore as Σ cannot contain both of these sentences the equation above writes X as a *disjoint union* of U_σ and $U_{\neg\sigma}$. Also, for two L-sentences σ and τ we have

$$U_\sigma \cap U_\tau = U_{\sigma\wedge\tau}$$

for $\text{Th}\,M \in U_\sigma \cap U_\tau$ holds if and only if both σ and τ are true in M, which holds if and only if $\sigma \wedge \tau$ is true in M. To define a topology on X we need to specify the collection of subsets of X to be called 'open', which must include $\varnothing$ and X and must be closed under finite intersections and arbitrary unions. We do this by saying that open sets are arbitrary unions of sets of the form U_σ. Since $U_\sigma \cap U_\tau = U_{\sigma\wedge\tau}$ these open sets are closed under finite intersection. The complements of these open sets, i.e. sets of the form $X \setminus U$ where U is open, are said to be *closed*.

One interesting property of our topology is that it is totally disconnected, which means that for any two distinct elements Σ and Γ of X, there are open sets U, V such that X is the disjoint union of U and V, and Σ is in one and Γ in the other. To see this is true, let σ be some first-order statement in one of Σ, Γ but not the other. (There must be such a σ as these sets are distinct.) Then we may take $U = U_\sigma$ and $V = U_{\neg\sigma}$, and the required properties hold.

Recall from Definition 8.26 that a topological space X is *compact* if whenever $\{V_i : i \in I\}$ is a collection of open subsets of X that covers X, there is a finite subcollection of the V_i that also covers X. Then, from this definition and the Compactness Theorem for first-order logic, our space X is compact.

Theorem 10.19 *The topological space X is compact.*

Proof Let $\{V_i : i \in I\}$ be a collection of open subsets of X, and suppose that no finite collection of the V_i covers X. We must show that there is $\Gamma \in X$ which is in none of the V_i.

We first reduce to the case when each V_i is of the form U_σ for some L-sentence σ. Since V_i is open and therefore the union of sets of the form U_σ, every element in V_i is contained in some U_σ. We now consider the collection of open sets $\{U_\sigma : U_\sigma \subseteq V_i$ for some $i \in I\}$ and note that it covers the same subset of X as $\{V_i : i \in I\}$, and also has the property that no finite subcollection of it can cover the whole of X. It suffices therefore to find $\Gamma \in X$ which is in none of the U_σ.

To this end, let

$$\Gamma = \{\neg \sigma : U_\sigma \subseteq V_i \text{ for some } i \in I\}.$$

We claim that there is an L-structure M_Γ in which all sentences in Γ hold. If not then, by the Compactness Theorem, there is a finite subset of Γ with no structure making it true:

$$\{\neg \sigma_1, \neg \sigma_2, \ldots, \neg \sigma_n\} \vDash \bot.$$

This means that no structure M satisfies all of $\neg \sigma_1, \neg \sigma_2, \ldots, \neg \sigma_n$ and hence no $\Sigma \in X$ can contain all of $\neg \sigma_1, \neg \sigma_2, \ldots, \neg \sigma_n$. Thus

$$U_{\sigma_1} \cup U_{\sigma_2} \cup \ldots \cup U_{\sigma_n} = X$$

since every structure makes one of $\sigma_1, \ldots, \sigma_n$ true. But this contradicts the assumption that no finite subcollection of the U_σ covers X. Hence there is a structure M_Γ in which Γ is true. Let $\Gamma^+ = \text{Th} M_\Gamma \in X$. It is now easy to see that Γ^+ is not an element of any of the U_σ used in the definition of Γ, since $\neg \sigma \in \Gamma \subseteq \Gamma^+$. This shows that the original collection of open sets U_σ does not cover X, as required. □

The reader may be interested to note that this argument only used the Compactness Theorem and the idea of an L-structure. Neither Soundness nor Completeness was needed at all. There is a converse argument too: the compactness of the space X implies the Compactness Theorem as previously given, since if Γ is a an infinite set of L-sentences with no structure satisfying all $\sigma \in \Gamma$ then

$$\{U_{\neg \sigma} : \sigma \in \Gamma\}$$

is an open cover of X since every M satisfies some $\neg \sigma$. Therefore there is a finite subcover

$$\{U_{\neg \sigma_i} : i = 1, 2, \ldots, n\}$$

and hence there is no L-structure satisfying all sentences in $\{\sigma_1, \sigma_2, \ldots, \sigma_n\}$.

Exercise 10.20 Express the Compactness Theorem for propositional logic as the compactness of a suitable topological space X. (For this, see also Proposition 8.35.)

10.4 The Omitting Types Theorem*

The Henkin method used in the proof of the Completeness Theorem is surprisingly powerful, and by inspecting it carefully we can get plenty of extra useful information, especially when the language is countable. Model theorists know this kind of argument as 'Omitting Types' for reasons that will become apparent, but from the point of view of topology (as discussed in Section 10.3) it is very close to the Baire Theorem for compact Hausdorff spaces.

The Omitting Types Theorem is an important and useful result in model theory, and many more advanced texts on model theory rightly give it a thorough presentation and a long discussion of its applications, how the proof works, and how the proof may be varied to give related results. Here I shall present only the very shortest of proofs for readers comfortable with the topological terminology used before. *Throughout this section we work with first-order logic in countable languages.*

We start by taking a countable first-order language L and a consistent set Σ_0 of L-sentences. We add a countably infinite set of constants, W (so the cardinality of W is the same as the cardinality of L) and denote by L_W the expanded language with these constant symbols added. We note the key fact that L_W is also a countable first-order language.

Now consider sets of L_W-sentences Σ that are *deductively closed*, i.e. for which

$$\Sigma \vdash \sigma \text{ implies } \sigma \in \Sigma$$

for each sentence σ of L_W. The set of all such Σ is called the *set of L_W-theories extending Σ_0* and will be denoted here by T. Thus

$$T = \{\Sigma \subseteq L_W : \Sigma \text{ is consistent, deductively closed and } \Sigma \supseteq \Sigma_0\}.$$

As in Section 10.3, we make T into a topological space by defining $U_\sigma = \{\Sigma \in T : \sigma \in \Sigma\}$ and saying that open sets are unions of sets of the form U_σ.

If X is a topological space, i.e. a set with specified open subsets, and $Y \subseteq X$ is a subset of X, then we can regard Y as a topological space with topology inherited from X by saying that a set $U \subseteq Y$ is open if and only if $U = Y \cap V$ for some open $V \subseteq X$. This topology on Y is called the *subspace topology*.

Exercise 10.21 In the case when $\Sigma_0 = \varnothing$, we can recover our space X of

Section 10.3 as a subspace of T. Show this as follows. First say that a set of L-sentences Σ is *complete for* L if $\Sigma \vdash \sigma$ or $\Sigma \vdash \neg \sigma$ for all sentences σ from L. Now prove that the set X of all $\mathrm{Th}\, M$ for some L-structure M is exactly the set of all consistent, deductively closed sets of L-sentences which are complete for L, and that the subspace topology induced on this by the topology on T is the same as the one given for X before.

Exercise 10.22 Prove that the space T is compact. (Hint: essentially the same proof as for X in the last section should work.)

In the proof of the Completeness Theorem, we focused on maximal elements of T, where maximal denotes with respect to set inclusion $\subseteq$. Let us denote the set of maximal elements of T as $T_{\max}$, and equip it with the subspace topology as before.

Theorem 10.23 *The space $T_{\max}$ is compact.*

Proof Consider a set C of sets of the form U_σ and suppose no finite subset of C covers $T_{\max}$. Let

$$\Gamma = \{\neg\sigma : U_\sigma \in C\}$$

and verify from our assumptions on C that $\Gamma \nvdash \bot$. Then Γ extends to a maximal $\Gamma' \supseteq \Gamma$, which is clearly deductively closed and in $T_{\max}$ but not in any $U_\sigma \in U$. ⊔

We met the idea of a totally disconnected space in Sections 8.4 and 10.3. *Totally disconnected spaces* are topological spaces X with the property that whenever $x \neq y$ in X then there are open sets U, V such that $x \in U$, $y \in V$, $U \cap V = \varnothing$ and $X = U \cup V$. (The sets U, V here are complements of each other and hence closed as well as open.) Totally disconnected spaces are extreme types of *Hausdorff spaces* which have the property that whenever $x \neq y$ in X then there are open sets U, V such that $x \in U$, $y \in V$ and $U \cap V = \varnothing$.

Proposition 10.24 *The space $T_{\max}$ is totally disconnected and hence Hausdorff.*

Proof Let $\Sigma, \Gamma \in T_{\max}$ be distinct. Then as these are distinct there is some $\sigma \in \Sigma$ that is not in Γ or some $\tau \in \Gamma$ that is not in Σ. Assume the first, so $\sigma \notin \Gamma$. Then by maximality of Γ we have $\neg\sigma \in \Gamma$ and hence $\Sigma \in U_\sigma$, $\Gamma \in U_{\neg\sigma}$ and $T_{\max} = U_\sigma \cup U_{\neg\sigma}$ is a disjoint union. □

In a small number of very special cases in certain languages L and base sets Σ_0 we can build an L-structure $M \vDash \Sigma^+$ directly from the set $\Sigma^+ \in T_{\max}$, often using an argument similar to the proof of the Completeness Theorem. Unfortunately, these arguments do not work in all cases and in general it requires the Henkin axioms to work. So we next abstract from the Henkin axioms those properties of a set $\Sigma^+ \in T_{\max}$ that are required for the construction in the Completeness Theorem to work.

Definition 10.25 Say that a set of L_W-sentences Σ has the *Henkin property* if whenever $\phi(x)$ is an L_W-formula in a single free variable then there is a constant $c \in W$ such that the sentence

$$\forall x \left(\neg \phi(x) \vee \phi(c) \right)$$

is Σ.

The Henkin property appears to be a rather technical property of a set of sentences in L_W. We are really interested in L_W-structures. The following lemma indicates the connection.

Lemma 10.26 *Suppose $\Sigma^+ \in T_{\max}$ has the Henkin property. Then there is $M \vDash \Sigma^+$ such that for all elements $a \in M$ there is some constant symbol $w \in W$ such that a realises w.*

Proof This is identical to the proof of the Completeness Theorem. We let $M = W/\sim$ where $u \sim v$ if and only if $u = v \in \Sigma^+$ and $M \vDash R([u_1], \ldots, [u_k])$ if and only if $R(u_1, \ldots, u_k) \in \Sigma^+$ for a k-ary relation symbol R, and similarly for functions. Then M is well defined and by an induction on formulas using the maximality of Σ^+ and the Henkin property we get that $M \vDash \Sigma^+$. □

So, not only do sets of sentences with the Henkin property have structures, but these structures can be chosen to consist entirely of constants from W.

By Lemma 10.4 and a Zorn's Lemma argument, maximal sets $\Sigma^+ \in T$ with the Henkin property exist, at least when we choose the constants W to be the Henkin constants in Lemma 10.4. But by renaming these constants, we see that maximal sets $\Sigma^+ \in T$ with the Henkin property exist whatever W is, provided it has the same cardinality as L. With the topological setting here we can review this argument and understand it in a more useful form.

First we need some more definitions and a result from topology.

Definition 10.27 Let X be a topological space and $A \subseteq X$. Then A is *co-rare* if whenever $U \subseteq X$ is open and non-empty then there is a non-empty open set

$V \subseteq A \cap U$. The set A is *co-meagre* if there are countably many co-rare sets A_i such that $A \supseteq \bigcap_{i=0}^{\infty} A_i$.

In the case of our space T where the topology is given by the sets U_σ, the definition of co-rare is easily seen to be equivalent to the following more convenient form: A is *co-rare* if for all non-empty U_σ there is a non-empty set $U_\tau \subseteq A \cap U_\sigma$.

Both of the notions 'co-rare' and 'co-meagre' just defined describe 'large' subsets. (There are corresponding notions of 'rare' and 'meagre' describing 'small' sets too.) In fact, the collection of co-meagre subsets $A \subseteq T$ is a filter in the boolean algebra $P(T)$ of all subsets of T, and this filter is closed under countable intersections. That this filter is proper is the content of the next result.

Theorem 10.28 (Baire) *Let X be a compact Hausdorff topological space and $A \subseteq X$ co-meagre. Then A is non-empty. In fact A is dense in X, meaning for any non-empty open $U \subseteq X$ the intersection $U \cap A$ is non-empty.*

Proof Suppose $A \supseteq \bigcap A_i$ where each A_i is co-rare and i ranges over elements of $\mathbb{N}$. Let $U_0 = U$, an arbitrary non-empty open set.

We now present an inductive construction of a sequence of non-empty open sets U_i. Assume we have U_i, where U_0 is as above, choose a non-empty open $V_i \subseteq A_i \cap U_i$ using the property that A_i is co-rare. Let $u_i \in V_i$. We find an open set U_{i+1} containing u_i and a closed set F_{i+1} such that $U_{i+1} \subseteq F_{i+1} \subseteq V_i$. To do this choose for each $x \notin V_i$ an open neighbourhood A_x of x and an open neighbourhood $B_x \subseteq V_i$ of u_i such that $A_x \cap B_x = \varnothing$. These neighbourhoods exist by the Hausdorff property of the space. The set V_i together with the collection of A_x forms an open cover of X, which is compact, and hence there is a finite subcover of X consisting of certain sets $V_i, A_{x_1}, \ldots, A_{x_k}$. Then let $U_{i+1} = \bigcap_j B_{x_j}$ and $F_{i+1} = \bigcup_j (X - A_{x_j})$. These have the required properties, as you may check.

The construction in the last paragraph gives a sequence of non-empty open sets U_i and closed sets F_i with $U_0 \supseteq F_1 \supseteq U_1 \supseteq F_2 \supseteq U_2 \supseteq \ldots$ and $U_{i+1} \subseteq A_i$ so $\bigcap U_i \subseteq A$. It suffices to show $\bigcap U_i \neq \varnothing$. But if this were not the case, the set $\bigcup_i (X \setminus F_i)$ would be an open cover of X without any finite subcover, contradicting compactness. $\qquad\Box$

In fact the last theorem can be generalised slightly. In particular it is true for locally compact Hausdorff spaces – full compactness is not required. This generalisation will not be required here, however.

The real power of Baire's Theorem is that it shows that any countable intersection of co-meagre sets is also co-meagre and hence non-empty. To see this, let A_i be co-meagre and $A_i \supseteq \bigcap A_{ij}$ where each A_{ij} is co-rare. But $\mathbb{N} \times \mathbb{N}$ is countable so there are in total a countable number of co-rare sets A_{ij} and hence $\bigcap_i A_i \supseteq \bigcap_{i,j} A_{ij}$ which is non-empty by the Baire Theorem.

The next theorem connects these ideas with logic.

Theorem 10.29 *The set of $\Sigma^+ \in T_{\max}$ with the Henkin property is a co-meagre subset of $T_{\max}$.*

Proof Consider a formula $\phi(x)$ in a single free variable x. To 'witness' this x with some $c \in W$ we need to look at the Henkin sentence $\forall x (\neg \phi(x) \vee \phi(c))$. The set of $\Sigma^+ \in T_{\max}$ that makes this statement true is $U_{\forall x (\neg \phi(x) \vee \phi(c))}$. In fact, we do not mind which particular constant witnesses x in $\phi(x)$. The set of $\Sigma^+ \in T_{\max}$ that witnesses x in $\phi(x)$ by some constant from W is the open set $H_{\phi(x)} = \bigcup_{c \in W} U_{\forall x (\neg \phi(x) \vee \phi(c))}$.

We stop a moment at this point to show that $H_{\phi(x)}$ is co-rare. Given a typical non-empty basic open set U_σ we need to consider $U_\sigma \cap H_{\phi(x)}$. This set is open; we need to show it is non-empty. Since U_σ is non-empty, $\Sigma_0 \cup \{\sigma\}$ is consistent. But the sentence σ can only mention finitely many constants $c \in W$. So some constant $d \in W$ is not mentioned in σ, and so by the lemma on Henkinisation (Lemma 10.4) $\sigma \wedge \forall x (\neg \phi(x) \vee \phi(d))$ is consistent. It follows by Zorn's Lemma that this sentence is contained in some maximal $\Sigma^+ \in T_{\max}$ and hence $\Sigma^+ \in H_{\phi(x)} \cap U_\sigma$, as required. Since this works for all U_σ, $H_{\phi(x)}$ is co-rare.

However, we are not interested in the set of Σ^+ that witness $\phi(x)$ for a *single* $\phi(x)$, but those that witness $\phi(x)$ for *all* $\phi(x)$. The set of these is

$$H = \bigcap_{\phi(x)} H_{\phi(x)}$$

which is a countable intersection of co-rare sets, hence co-meagre, since the set of all $\phi(x)$ is countable as L_W is a countable language. This completes the proof $\square$

As mentioned, any countable intersection of co-meagre sets is co-meagre and hence non-empty. This suggests that we should find other co-meagre sets to intersect with the set of Σ^+ having the Henkin property.

Definition 10.30 Let p be a set of L-formulas $\phi(x)$ in a single free variable x. Say that a set of L_W-sentences Σ omits p if whenever $c \in W$ is a constant symbol then there is a formula $\phi(x) \in p$ such that $\neg \phi(c) \in \Sigma$.

We are going to find $\Sigma \in T_{\max}$ with the Henkin property that omits one or more sets p. However we will not be able to omit every set p. For example, when Σ_0 is a theory of numbers and p is the set of all properties $\phi(x)$ true of a prime x, then presumably p cannot be omitted as Σ_0 already implies that the prime number x exists. More subtly, constructions using the Baire theorem are usually complex affairs with many parts happening 'simultaneously' and we may not be able to omit p because some other part of the construction may have made some statement $\exists x \, \psi(x)$ true and p is a set of properties true of any x satisfying $\psi(x)$. We define the class of p we can hope to omit in the following definition.

Definition 10.31 Suppose p is a set of L-formulas $\phi(x)$ in a single free variable x, and Σ_0 is a set of L-sentences. Say that p is *isolated* over Σ_0 if there is some formula $\psi(x)$ of L such that $\Sigma_0 \cup \{\exists x \, \psi(x)\} \not\vdash \bot$ and $\Sigma_0 \vdash \forall x \, (\psi(x) \rightarrow \phi(x))$ for all $\phi(x) \in p$. Such a formula $\psi(x)$ as here is said to be a *support* of p.

Lemma 10.32 *Suppose p is a countable set of L-formulas with no support over Σ_0. Then the set of $\Sigma^+ \in T_{\max}$ which omits p is a co-meagre subset of $T_{\max}$.*

Proof Start by looking at a single $c \in W$. We want this c to fail to satisfy all properties in p. That is we want some $\neg \phi(c)$ to be true. That tells us to look at the set

$$S_{p,c} = \bigcup\nolimits_{\phi(x) \in p} U_{\neg \phi(c)}.$$

The set of Σ^+ in S_c are those that omit the set p at c. Let us show that S_c is co-rare. It is clearly open, so let σ be a sentence of L_W and suppose U_σ is non-empty. The sentence σ might involve c, so we write it as $\sigma(c)$. However the sentence $\exists x \, \sigma(x)$ is not a support of p, by our assumptions on p. So, as $\Sigma_0 \cup \{\exists x \, \sigma(x)\}$ is consistent, by the definition of support there must be some $\phi(x) \in p$ such that $\Sigma_0 \{\exists x \, (\sigma(x) \wedge \neg \phi(x))\}$ is consistent. But that means that $U_{\sigma(c) \wedge \neg \phi(c)} = U_{\sigma(c)} \cap U_{\neg \phi(c)}$ is non-empty, as required. The lemma now follows as the set of Σ^+ omitting p is

$$S_p = \bigcap\nolimits_{c \in W} S_{p,c}$$

which is a countable intersection of co-rare sets and hence is co-meagre. $\quad\square$

Similar considerations apply to sets of first-order properties of a collection of variables.

Definition 10.33 Suppose p is a set of L-formulas $\phi(x_1, \ldots, x_k)$ in free variables $x_1, \ldots, x_k$, and Σ_0 is a set of L-sentences. Then p is *isolated* over Σ_0 if there is some formula $\psi(x_1, \ldots, x_k)$ of L such that $\Sigma_0 \cup \{\exists x \, \psi(x_1, \ldots, x_k)\} \nvDash \bot$ and $\Sigma_0 \vdash \forall x \, (\psi(x_1, \ldots, x_k) \to \phi(x_1, \ldots, x_k))$ for all $\phi(x_1, \ldots, x_k) \in p$. Such a formula $\psi(x_1, \ldots, x_k)$ is said to be a *support* of p. A set of L_W-sentences Σ *omits* p if whenever $c_1, \ldots, c_k \in W$ are constants from W then there is $\phi(x_1, \ldots, x_k) \in p$ such that $\neg \phi(c_1, \ldots, c_k) \in \Sigma$.

Lemma 10.34 *Suppose p is a countable set of L-formulas in free variables $x_1, \ldots, x_k$ with no support over Σ_0. Then the set of $\Sigma^+ \in T_{\max}$ which omits p is a co-meagre subset of $T_{\max}$.*

Proof Identical to the previous proof, but using

$$S_{p, c_1, \ldots, c_k} = \bigcup\nolimits_{\phi(x_1, \ldots, x_k) \in p} U_{\neg \phi(c_1, \ldots, c_k)}$$

and the fact that the set W^k of k-tuples $c_1, \ldots, c_k$ of constants from W is also countable. $\qquad\square$

The reader can probably guess by now the terminology that is typically used in situations like this. A set p of formulas of the form $\phi(x_1, \ldots, x_k)$ is often called a 'type'. (It represents a set of properties that may be held by some $a_1, \ldots, a_k \in M$, and hence represents a 'kind' or 'type' of k-tuple from M.) Also, a type is *omitted* by M if no tuple $a_1, \ldots, a_k \in M$ has the property represented by p.

If a type is not omitted in a structure M it is said to be *realised* in M. In general, omitting a type is the more difficult thing to do: realising a type simply requires the Compactness Theorem, as the next exercise shows.

Exercise 10.35 Let Σ_0 be a consistent set of L-sentences, where L is a countable first-order language, and for each n let p be a set of L-formulas of the form $\phi(x_1, \ldots, x_{k_n})$ in free variables $x_1, \ldots, x_{k_n}$. Suppose that each p_n is *finitely satisfiable* over Σ_0, that is, for each finite $q \subseteq p_n$ there is $M_q \vDash \Sigma_0$ with some $a_1, \ldots, a_{k_n} \in M_q$ which satisfy every $\phi(x_1, \ldots, x_{k_n})$ in q. Using the Compactness Theorem show that there is a single $M \vDash \Sigma_0$ in which none of the p_n is omitted.

In contrast, the Omitting Types Theorem says that isolated types can be omitted. All that remains is to state this formally and put all the ingredients above together to give the proof.

Theorem 10.36 (Omitting Types) *Let Σ_0 be a consistent set of L-sentences, where L is a countable first-order language, and for each $n \in \mathbb{N}$ suppose that*

p_n is a set of L-formulas $\phi(x_1, \ldots, x_{k_n})$ in free variables $x_1, \ldots, x_{k_n}$ and p_n is isolated over Σ_0. Then there is a countable L-structure M such that $M \vDash \Sigma_0$ and in which each p_n is omitted.

Proof We will find a maximal $\Sigma^+ \in T_{\max}$ with the Henkin property. By Lemma 10.26 this gives an L_W-structure $M \vDash \Sigma^+$ in which every element is a constant from W. If in addition Σ^+ omits each p_n then each type p_n is omitted by each tuple of constants, and hence is omitted in M.

   Thus we must find

$$\Sigma^+ \in H \cap \bigcap_{n \in \mathbb{N}} S_{p_n}$$

where H is the set of Σ^+ with the Henkin property, in Theorem 10.29, and S_{p_n} is the set of Σ^+ that omits p_n from Lemma 10.34. But this is a countable intersection of co-meagre sets, hence is non-empty by the Baire Theorem. Thus some Σ^+ exists and we have proved the theorem.   □

11

Model theory

11.1 Countable models and beyond

Model theory is the study of arbitrary *L*-structures for first-order languages *L*. It is a sort of generalised algebraic theory of algebraic structures. We talk of a structure *M* being a *model* of a set of *L*-sentences Σ when $M \vDash \Sigma$, and this gives the name to the theory. Actually, in practice, model theory tends to be much more about the structures themselves and the subsets and functions that are definable in those structures by first-order formulas, and much less about first-order sentences, but the term 'model theory' seems to be fixed now. This chapter attempts to give a flavour of model theory and presents some of the first theorems. It also contains a considerable amount of preliminary material on countable sets and cardinalities that we have somehow managed to put off until now.

Model theory starts off with the Compactness Theorem for first-order logic, which is phrased entirely in terms of the notion of semantics, $\vDash$, but was proved in the last chapter by an excursion into the realm of formal proofs. The key result guaranteeing the existence of models of a set of sentences Σ is the Completeness Theorem for first-order logic which provides us with a model *M* of Σ, under the assumption that $\Sigma \nvdash \bot$. We will start by looking at the Completeness Theorem in a little more detail to see what extra it can say for us.

The construction of the model *M* that the proof of the Completeness Theorem gives us is rather mysterious, as it relies on Zorn's Lemma to do the work of finding the maximal set extending Σ, and aside from choosing the sentences in Σ in the first place, we have little input. It is natural to ask a little more about this structure *M* we end up with.

As it happens, we cannot really say very much more about it: most questions about *M* that can be settled at all are answered by looking at the initial data Σ. In other words, almost all we know about *M* is the Σ that we started with.

160

If you need a particular M with special properties it is usually necessary to start with a different Σ (possibly in a different language) or, for more advanced work, apply more sophisticated variations of the Henkin construction given above such as the Omitting Types Theorem of Section 10.4. One of the few exceptions to this rule is the 'size' or cardinality of M. This is what we shall start our study of model theory with here.

We shall start with the countable case, and then explain how similar results apply in other infinite cases. As the discussion of uncountable structures inevitably involves some set theoretic background, the key technical results will be stated here but proved in an optional section, later on in this chapter.

The idea of a countable set was introduced earlier, in Definition 2.13. Our definition said that a set X is countable if it is empty or if there is a surjection from $\mathbb{N}$ to X. There is an alternative that is also convenient at times, indicated in the following proposition.

Proposition 11.1 *A set X is countable if and only if there is an injection* $f\colon X \to \mathbb{N}$.

Proof Suppose X is countable. Then if X is empty, by convention the empty function f from X to $\mathbb{N}$ is a perfectly good injection. (We do not have to specify any values $f(x)$ because there are none.) Otherwise, suppose $g\colon \mathbb{N} \to X$ is a surjection. Then we can define $f\colon X \to \mathbb{N}$ by letting $f(x)$ be the least $n \in \mathbb{N}$ such that $g(n) = x$. This is well defined as g is onto X, and is clearly an injection as no two distinct $x, y \in X$ can receive the same n since this would mean $x = g(n) = y$.

Conversely, if $f\colon X \to \mathbb{N}$ is an injection, then if X is empty it is countable, so suppose not. Choose $x_0 \in X$ and let $g(n) = x_0$ if $n \notin \operatorname{im}(f)$ and let $g(n)$ be the unique $x \in X$ such that $g(x) = n$ otherwise. It is easy to check that this defines a surjection $g\colon \mathbb{N} \to X$. $\qquad\square$

We are mainly interested in infinite countable sets, also called *countably infinite* sets. The canonical example of such a set is the set $\mathbb{N}$ of natural numbers. However, many other such sets are countable, and there are some surprises here too. We start with some reasonably pleasant consequences of our notion of 'countable'.

Proposition 11.2 *Let A be a countable set and $f\colon A \to B$ a bijection. Then B is countable.*

Proof If A is empty, then B must be too, for otherwise f cannot be surjective.

And if $g: \mathbb{N} \to A$ is a surjection then the composition of g and f is a surjection $\mathbb{N} \to B$. $\square$

Proposition 11.3 *Any subset of a countable set is countable.*

Proof If A is countable and $f: A \to \mathbb{N}$ is an injection, then for each $B \subseteq A$ the restriction of f to B is also an injection $B \to \mathbb{N}$. $\square$

Proposition 11.4 *If A is a countable set and $\sim$ is an equivalence relation on A then the set $A/\sim$ of equivalence classes of A is countable.*

Proof Suppose $f: \mathbb{N} \to A$ is a surjection and define $g: A \to A/\sim$ by $g(a) = a/\sim$, the equivalence class of a. Then the composition of f and g is a surjection $\mathbb{N} \to A/\sim$. $\square$

Proposition 11.5 *If A and B are countably infinite sets then there is a bijection $f: A \to B$.*

Proof We suppose A is countably infinite and show that there is a bijection $g: \mathbb{N} \to A$. The result will follow from a similar fact about B and by composing bijections.

So suppose $h: A \to \mathbb{N}$ is an injection. Then $\mathrm{im}(h)$ cannot be empty nor have a largest element n, for if it did, A would have at most $n + 1$ elements and would be finite. Thus $\mathrm{im}(h)$ contains arbitrarily large natural numbers and there is a function $k: \mathbb{N} \to \mathbb{N}$ such that $k(0)$ is the first element of $\mathrm{im}(h)$, $k(1)$ is the second element of $\mathrm{im}(h)$, ..., and $k(n)$ is the $n + 1$st element of $\mathrm{im}(h)$. (This function k is defined inductively by $k(0)$ is the least element of $\mathrm{im}(h)$ and $k(n+1)$ is the least element of $\mathrm{im}(h) \cap \{m \in \mathbb{N}: m > k(n)\}$.) Then the function $g: \mathbb{N} \to A$ defined by $g(n) = h^{-1}(h(n))$ is a bijection. $\square$

The surprise comes next.

Proposition 11.6 *If A, B are countable sets then their union $A \cup B$ and cartesian product $A \times B$ are both countable.*

Proof Suppose $f: A \to \mathbb{N}$ and $g: B \to \mathbb{N}$ are injections.

We can define an injection $h: A \cup B \to \mathbb{N}$ by $h(x) = 2f(x)$, if $x \in A$, and $h(x) = 2g(x) + 1$, if $x \notin A$. It is easy to check this is also an injection, hence $A \cup B$ is countable.

For the cartesian product, note that the function $p: \mathbb{N} \times \mathbb{N} \to \mathbb{N}$ given by

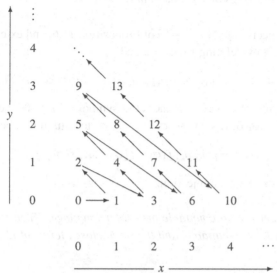

Figure 11.1 A pairing function for the natural numbers.

$p(x, y) = (x+y)(x+y+1)/2 + y$ is in fact a bijection pairing up two natural numbers as a single number. This is because $(x+y)(x+y+1)/2$ gives the number of points $(r, s) \in \mathbb{N} \times \mathbb{N}$ for which $r + s < x + y$ and thus the pairs mapping to values between $k(k+1)/2$ and $(k+1)(k+2)/2 - 1$ are the pairs $(x, y) \in \mathbb{N} \times \mathbb{N}$ with $x+y = k$ (see Figure 11.1). Thus we can define an injection $k : A \times B \to \mathbb{N}$ by $k(a, b) = p(f(a), g(b))$, showing $A \times B$ is countable. ☐

The first time you see this, it is quite startling news. For example, this proposition implies that $\mathbb{Z}$ and $\mathbb{Q}$ are countable. $\mathbb{Z}$ is countable because it is the disjoint union of two copies of $\mathbb{N}$ (namely the non-negative and the negative integers) and $\mathbb{Q}$ is countable since it can be constructed as a quotient of the set $\mathbb{Z} \times (\mathbb{Z} \setminus \{0\})$. The countability of the set of $\mathbb{Z}$ and $\mathbb{Q}$ seems to be counter-intuitive as $\mathbb{Q}$ contains so many more elements than $\mathbb{N}$, but as we can now see $\mathbb{Q}$ and $\mathbb{N}$ turn out to be exactly the same size.

These simple propositions on countability are enough to sharpen the Completeness Theorem quite considerably.

Definition 11.7 A *countable first-order language* is a first-order language L in which the set of variables, the set of constants, the set of relation symbols, and the set of function symbols, are all countable.

Proposition 11.8 *Let L be a countable first-order language. Then there are countably many finite strings of symbols from L.*

Proof We use the bijection $p: \mathbb{N} \times \mathbb{N} \to \mathbb{N}$ of Proposition 11.6, and extend it to bijections $p_k: \mathbb{N}^k \to \mathbb{N}$ by defining $p_1(n) = n$ and

$$p_{k+1}(n_1, n_2, \ldots, n_k, n_{k+1}) = p(p_k(n_1, n_2, \ldots, n_k), n_{k+1}).$$

Now given an injection $f: L \to \mathbb{N}$ from the set of symbols of L to $\mathbb{N}$, we define g from strings $\sigma_1\sigma_2\ldots\sigma_k$ of L-symbols $\sigma_1, \sigma_2, \ldots, \sigma_k$ to natural numbers by

$$g(\sigma_1\sigma_2\ldots\sigma_k) = p(k, p_k(f(\sigma_1), f(\sigma_2), \ldots, f(\sigma_k)))$$

and it is easy to check this is an injection.   □

Proposition 11.9 *Let L be a countable first-order language. Then the complete Henkinisation L^H is countable, and the set of closed terms of L^H is also countable.*

Proof It suffices, by the previous proposition, to show that there are countably many symbols in L^H, for the set of closed terms in this language is a set of strings from these symbols.

Now there are countably many strings of the form $\phi(x)$ from L, and so the first Henkinisation $H(L)$ consists of a union of a countable set of symbols L and an additional countable set $\{\varepsilon_{\phi(x)} : \phi(x) \text{ from } L\}$. So $H(L)$ is countable by Proposition 11.6, and what is more this process gives a completely definite method for obtaining an injection $H(f): H(L) \to \mathbb{N}$ from an injection $f: L \to \mathbb{N}$.

To define an injection $L^H \to \mathbb{N}$, take a symbol s of L^H and let $n \in \mathbb{N}$ be least such that s is in the nth Henkinisation $H^n(L)$ of L. Then we define $g(s) = p(n, H^n(s))$, the number for the pair consisting of n and the number for the symbol s in $H^n(s)$.   □

Definition 11.10 An L-structure M is said to be *countable* if its domain or underlying set is countable.

Theorem 11.11 (Completeness Theorem, countable version) *Let Σ be a set of sentences in a countable first-order language L, and suppose that $\Sigma \not\vdash \bot$. Then there is a countable model of Σ.*

Proof Just repeat the previous proof of the Completeness Theorem, noting that the structure obtained has domain which is a quotient of the set of closed terms of the Henkinised language, so is countable by Proposition 11.9 and Proposition 11.4. □

This new version of the Completeness Theorem is quite striking too. For example, consider the structure $\mathbb{R}$ consisting of the real numbers together with the constants 0, 1, the functions $+$, $\times$, $-$, and the relation $<$. The corresponding first-order language L is countable. Now let Σ be the set of all L-sentences true in $\mathbb{R}$. By the Soundness Theorem $\Sigma \not\vdash \bot$, so by the countable form of the Completeness Theorem Σ has a countable model. This cannot be $\mathbb{R}$ itself, as $\mathbb{R}$ is uncountable.

Even more disconcertingly, it is possible to express all our axioms of set theory, the Zermelo–Fränkel theory of sets, in a countable first-order language. In Zermelo–Fränkel set theory it is possible to discuss and prove all of our results on countability and uncountability – including the uncountability of $\mathbb{R}$. It follows that there is a countable model M of set theory, with its own countable version of $\mathbb{R}$, that makes the statement '$\mathbb{R}$ is uncountable' true. This paradox is known as 'Skolem's paradox'. It is resolved by observing that the notions of 'countable' in our M and in the real world are actually rather different. Rather than concluding that we cannot in fact talk about countable and uncountable sets in any coherent way, it is perhaps better to conclude that a countable first-order language is not actually expressive enough to capture our notion of countability in the real world correctly.

We now turn to other infinite sets, such as the set of real numbers, which was shown to be uncountable by Cantor (see Exercise 2.28). We will need some results on the sizes of these sets analogous to our results on countable sets, and as these results are really 'set theory' rather than 'logic' we shall not prove them here. They can all be proved from Zorn's Lemma, and proofs are given in an optional section in this chapter. The size of a set will be called its *cardinality*, and we will need to compare cardinalities, saying when two cardinalities are equal, or when one is larger than the other. The next two definitions explain this.

Definition 11.12 Two sets A and B have the *same cardinality* if there is a bijection $f \colon A \to B$.

Definition 11.13 Let A and B be sets. We write $\operatorname{card} A \leqslant \operatorname{card} B$ to mean there is an injection $f \colon A \to B$. We write $\operatorname{card} A < \operatorname{card} B$ to mean there is an injection $f \colon A \to B$ but no bijection $g \colon A \to B$.

A very special case of this is that of 0, the cardinality of the empty set: this is less than or equal to all other cardinalities, since there is always an injection from $\varnothing$ to any other set, the empty function, as discussed earlier.

Remark 11.14 Definitions 11.12 and 11.13 beg a difficult question: what mathematical object *is* the cardinality of a set? In fact nothing later on will require a specific answer to this question, so unless you are curious, this remark can be safely skipped. But there are perfectly reasonable definitions of 'cardinal'. One attempt goes as follows. As the relation of 'having the same cardinality' is an equivalence relation on the class of all sets, the cardinality $\operatorname{card} A$ of a set A should be the equivalence class of A under this relation. That will be good enough for us here, but it is not quite right since – as it happens – this equivalence class is not a set in the official sense of Zermelo–Fränkel set theory. A technical variation of this definition (known as Scott's trick) does succeed in defining a set $\operatorname{card} A$ representing the cardinality of A as the equivalence class of all sets in a specially chosen sub-universe of sets containing sets of the same cardinality as A. With the Axiom of Choice there are several more attractive alternative definitions of $\operatorname{card} A$, including defining it to be the least ordinal in one-to-one correspondence with A, so the problem can be resolved, though not without going somewhat beyond the scope of this book.

Definition 11.15 For a finite set A with n elements we let $\operatorname{card} A = n$ and, following the notation first given by Cantor, if A is a countably infinite set we write $\aleph_0$ for the cardinality $\operatorname{card} A$, where $\aleph$ is the first Hebrew letter 'aleph'.

The next three propositions explore the $\leqslant$ relation further; they do not seem very surprising, but should not be belittled. None of these propositions is obvious or straightforward to prove. Proofs appear later in the optional section.

Proposition 11.16 *Suppose there is a surjection* $f \colon A \to B$. *Then* $\operatorname{card} B \leqslant \operatorname{card} A$, *i.e. there is an injection* $g \colon B \to A$.

Proposition 11.17 (Schröder–Bernstein Theorem) *Let A and B be sets and suppose that* $\operatorname{card} A \leqslant \operatorname{card} B$ *and* $\operatorname{card} B \leqslant \operatorname{card} A$; *then* $\operatorname{card} A = \operatorname{card} B$. *In other words if there are injections $A \to B$ and $B \to A$ then there is a bijection $A \to B$.*

Proposition 11.18 (Trichotomy Theorem) *Let A and B be sets. Then exactly one of the following is true:* $\operatorname{card} A < \operatorname{card} B$; $\operatorname{card} A = \operatorname{card} B$; *or* $\operatorname{card} B < \operatorname{card} A$.

There are results analogous to earlier propositions on countable sets for arbitrary cardinalities. For example, if $A \subseteq B$ then $\operatorname{card} A \leqslant \operatorname{card} B$ since the map $a \mapsto a$ is an injection $A \to B$. If $\sim$ is an equivalence relation on A then $\operatorname{card}(A/\sim) \leqslant \operatorname{card} A$ as the map $a \mapsto a/\sim$ is a surjection $A \to A/\sim$ and we may apply Proposition 11.16 to obtain an injection $A/\sim \to A$. Here is another application of the above results.

Proposition 11.19 *Let A be infinite. Then $\aleph_0 \leqslant \operatorname{card} A$, i.e. there is an injection* $\mathbb{N} \to A$.

Proof By trichotomy, if the conclusion is false then $\operatorname{card} \mathbb{N} > \operatorname{card} A$. This would mean that there is an injection $f: A \to \mathbb{N}$ but no bijection between A and $\mathbb{N}$. This is impossible since A is infinite, so the image $\operatorname{im} f \subseteq \mathbb{N}$ is also infinite and any infinite subset of $\mathbb{N}$ is in one-to-one correspondence with the whole of $\mathbb{N}$, by Proposition 11.5, which would mean A is also in one-to-one correspondence with $\mathbb{N}$. $\qquad\square$

We are interested in uncountable versions of the Completeness Theorem. For that we need to compute the number of symbols in the complete Henkinisation L^H of an uncountable language L. To do this we will need to add and multiply infinite cardinalities.

Definition 11.20 Let A, B be sets, and $\kappa = \operatorname{card} A$, $\lambda = \operatorname{card} B$. Then

   (i) $\kappa + \lambda$ is the cardinality of the set $(A \times \{0\}) \cup (B \times \{1\})$,

   (ii) $\kappa\lambda$ is the cardinality of the set $A \times B$.

Proposition 11.21 *The cardinal arithmetic operations of addition and multiplication are well defined. That is, they do not depend on the choice of sets A, B.*

Proof Suppose $\operatorname{card} A = \operatorname{card} A'$ and $\operatorname{card} B = \operatorname{card} B'$, so there are bijections $f: A \to A'$ and $g: B \to B'$. Then

$$h: (A \times \{0\}) \cup (B \times \{1\}) \to (A' \times \{0\}) \cup (B' \times \{1\})$$

defined by

$$h(a, 0) = (f(a), 0) \in A' \times \{0\} \text{ and } h(b, 1) = (g(b), 1) \in B' \times \{1\}$$

for $a \in A$ and $b \in B$, is a bijection, since f, g are. (Exercise: verify this.) Also,

$$k: A \times B \to A' \times B'$$

defined by

$$k(a, b) = (f(a), g(b)) \in A' \times B'$$

is also a bijection. Thus $\text{card}(A \times \{0\} \cup B \times \{1\}) = \text{card}(A' \times \{0\} \cup B' \times \{1\})$ and $\text{card}(A \times B) = \text{card}(A' \times B')$. □

Other natural properties of cardinal arithmetic such as associativity, commutativity and distributivity also work as you might expect. For infinite cardinalities, these operations are quite straightforward, and the next proposition is our key result for calculating the cardinality of L^H. Again, it is not at all easy, and a proof from Zorn's Lemma will be found in an optional section later on.

Theorem 11.22 (Cardinal Arithmetic) *Let A, B be infinite sets, and* $\kappa = \text{card}\,A$, $\lambda = \text{card}\,B$. *Then* $\kappa + \lambda = \kappa\lambda = \max\{\kappa, \lambda\}$.

With all these preliminaries out of the way, we can get back to logic. Since proofs are finite objects, there is never any need for more than countably many variable symbols in a first-order language, so from now on we shall assume all of our first-order languages have countably many variables.

Definition 11.23 Let L be a first-order language. Then the *cardinality* $\text{card}\,L$ of L is the cardinality of the set of symbols in L. In other words, using Theorem 11.22, $\text{card}\,L$ is the maximum of: (a) $\aleph_0$; (b) the cardinality of the set of constant symbols of L; (c) the cardinality of the set of relation symbols of L; (d) the cardinality of the set of function symbols of L.

Proposition 11.24 *Let L be a first-order language. Then the cardinality of the set of strings of symbols from L equals* $\text{card}\,L$

Proof This is similar to the countable case, but using the proposition on cardinal arithmetic instead of a directly constructed pairing function.

Let S be the set of symbols from L, so $\text{card}\,S = \kappa$ is infinite. By cardinal arithmetic (Theorem 11.22) there is a bijection $p\colon S \times S \to S$. We extend this to $p_k\colon S^k \to S$ by $p_1(s) = s$ and

$$p_{k+1}(n_1, n_2, \ldots, n_k, n_{k+1}) = p(p_k(n_1, n_2, \ldots, n_k), n_{k+1}).$$

Also, as S is infinite, there is an injection $q\colon \mathbb{N} \to S$, by Proposition 11.19. So we may define g from strings of symbols of L to S by

$$g(\sigma_1\sigma_2\ldots\sigma_k) = p(q(k), p_k(\sigma_1, \sigma_2, \ldots, \sigma_k))$$

and this is readily seen to be an injection. □

Proposition 11.25 *Let L be a first-order language. Then the cardinality of the complete Henkinisation L^H of L is the same as that of L; the cardinality of the set of closed terms of L^H is also equal to* $\mathrm{card}\,L$.

Proof This proof is almost identical to the countable case, Proposition 11.9. Let $\kappa = \mathrm{card}\,L$. It suffices to show that there are κ many symbols in L^H.

Now there are κ many strings of the form $\phi(x)$ from L, and so the first Henkinisation $H(L)$ consists of a disjoint union of L with a set of constant symbols $\{\varepsilon_{\phi(x)} : \phi(x) \text{ from } L\}$ of cardinality κ. So $\mathrm{card}\,H(L) = \kappa + \kappa = \kappa$ by cardinal arithmetic, and this gives a bijection $h_1 : H(L) \to L$ from the symbols of $H(L)$ to the symbols of L. Similarly we can choose for each $k \in \mathbb{N}$ a bijection $h_k : H^k(L) \to L$ from the symbols of the kth Henkinisation of L to the symbols of L. We also have, by Proposition 11.19 and cardinal arithmetic, an injection $p : \mathbb{N} \times L \to L$.

Now, to define an injection $L^H \to L$, take a symbol s of L^H and let $k \in \mathbb{N}$ be least such that s is in the kth Henkinisation $H^k(L)$ of L. Then define a map $g(s) = p(k, h_k(s))$ taking the symbol s to a symbol in L 'encoding' the pair consisting of k and the symbol s. This is readily seen to be an injection. $\square$

Definition 11.26 The *cardinality of* an L-structure is the cardinality of its underlying set, i.e. the cardinality of the non-empty domain of elements of the structure.

Theorem 11.27 (Completeness Theorem, arbitrary cardinalities) *Let Σ be a set of sentences in a first-order language L, and suppose that $\Sigma \not\vdash \bot$. Then there is a model of Σ of cardinality at most* $\mathrm{card}\,L$.

Proof Just repeat the previous roof, noting that the domain of the structure it gives is a quotient of the set of closed terms of the Henkinised language, and therefore has cardinality at most this. $\square$

It does not take very much work to modify this result to obtain models of all possible infinite cardinalities. All we need is an additional assumption that our set Σ has some infinite model. More specifically, we use the following theorem.

Theorem 11.28 *Let Σ be a consistent set of sentences in a first-order language L, let $\kappa \geqslant \mathrm{card}\,L$, and suppose that Σ has at least one infinite model. Then Σ has an infinite model of cardinality κ.*

Proof Let C_κ be a set of constants of cardinality κ and let L^+ be the language L with these constant symbols added. Then $\operatorname{card} L^+$ is exactly κ, by cardinal arithmetic. Also, let T_κ be the set of all statements $\neg c = d$ for $c \neq d \in C_\kappa$. Observe that $T_\kappa \cup \Sigma$ is consistent. This is because in any infinite model M of Σ, any finite subset of constants from C_κ can be interpreted in M in a way making them all different.

It follows by the Completeness Theorem that there is a model $M_\kappa \vDash T_\kappa \cup \Sigma$ of cardinality at most κ. That is, there is an injection $M \to C$, where C is our set of new constants which has cardinality κ. But by the statements T_κ, the map $c \mapsto [c]$ taking the constant symbol $c \in C$ to the element of M realising it, i.e. to the equivalence class of c, is also an injection $C \to M$. It follows by the Schröder–Bernstein Theorem that there is a bijection $C \to M$ and hence $\operatorname{card} M = \operatorname{card} C$. By discarding the additional constants in C_κ we obtain an L-structure of cardinality κ making Σ true. $\qquad\qquad\square$

At last, we can do some model theory and give an application of these ideas to mathematical theories formalised in first-order languages. First, we need a definition.

Definition 11.29 Let L be a first-order language and Σ be a set of L-sentences. We say that Σ *is complete (for L)* if whenever σ is an L-sentence we have either $\Sigma \vdash \sigma$ or $\Sigma \vdash \neg \sigma$.

This should not be confused with the Completeness Theorem. The Completeness Theorem says that our *logic* is complete, i.e. can prove all valid assertions. Think of Σ as a mathematical theory, such as a theory of groups, formalised in a first-order language L using first-order logic. Then Σ is complete if it can decide all mathematical statements that can be expressed in L, for example our theory of groups is complete if it decides all first-order statements about groups.

Example 11.30 Let L be the first-order language with the usual logical symbols, including $=$, but with no additional logical symbols. Let σ_n be an L-sentence expressing the fact that there are at least n elements (see Exercise 10.7). Let $\Sigma = \{\sigma_n : n \in \mathbb{N}\}$. Then it turns out that Σ is complete, i.e. for every σ in L either $\Sigma \vdash \sigma$ or $\Sigma \vdash \neg \sigma$. (We will prove this in a moment.) On the other hand, no finite subset Σ_0 of Σ is complete since, given Σ_0 there is always some $n \in \mathbb{N}$ such that $\Sigma_0 \nvdash \sigma_n$ and $\Sigma_0 \nvdash \neg \sigma_n$ as we may take models of any particular finite cardinality above some minimum cardinality determined by Σ_0.

Complete sets Σ are of interest for many reasons, not least in the study of the foundations of mathematics, where complete sets provide examples of mathematical theories where all mathematical statements (in the appropriate language) can be determined true or false by a formal proof. It was a sort of Holy Grail of logic in the early twentieth century to find a proof system for the whole of mathematics, or a large part of it such as number theory or set theory, that was complete for its language. Such a system would, in principle, put mathematicians out of work, to be replaced by a computer churning out theorems in the system. That this Grail is in fact unattainable was eventually proved by Gödel in 1931. However, in looking for such theories a number of people, such as Tarski and Hilbert, found interesting first-order systems that were complete for their (limited) languages.

In a way exactly analogous to isomorphisms in traditional algebraic contexts, we say that two models M, N for the same language L are *isomorphic* if there is a bijection from M to N that preserves the L-structure on M, N. Such isomorphisms necessarily preserve the truth of L-sentences. (A formal proof of this fact is an induction on the number of symbols in a formula, and is left as an exercise for the pedantic reader.) We now use this idea to give another important model theoretic notion.

Definition 11.31 Let L be a first-order language, Σ be a set of L-sentences, and κ an infinite cardinal. We say Σ *is* κ-*categorical* if all models of Σ of cardinality κ are isomorphic to each other.

Theorem 11.32 *Let L be a first-order language, and Σ be a set of L-sentences. Suppose Σ is κ-categorical for some infinite cardinal $\kappa \geqslant \operatorname{card} L$ and Σ has no finite models. Then Σ is complete.*

Proof If Σ is not complete there is an L-sentence such that both $\Sigma_1 = \Sigma \cup \{\sigma\}$ and $\Sigma_2 = \Sigma \cup \{\neg \sigma\}$ are consistent. Since Σ has no finite models, by Theorem 11.28 there are models $M_1 \vDash \Sigma_1$ and $M_2 \vDash \Sigma_2$ of cardinality κ. But these models cannot be isomorphic since they do not agree about the truth of the sentence σ, so Σ is not κ-categorical. $\qquad\square$

Example 11.33 The set Σ of Example 11.30 is complete, as any two countably infinite models for the language with no non-logical symbols are necessarily isomorphic, and Σ has no finite models.

Theorem 11.32 gives interesting information in the following example.

Example 11.34 Let F be a field, such as $\mathbb{R}$, and consider infinite vector spaces over F. We will present a first-order language L_F for such vector spaces. L_F has binary function $+$ for addition of vectors, unary function $-$ for additive inverses, and a constant symbol 0 for the zero vector. It also has, for each $\lambda \in F$, a unary function s_λ for scalar multiplication by λ. It is possible to write down a set Σ_F of first-order axioms for F-vector spaces in this language (Exercise 11.37). Note that our language works for a fixed F only, and there are no symbols for addition or multiplication in F.

Then Σ_F is complete, since there are models of Σ_F of every infinite cardinality $\kappa \geqslant \mathrm{card}\,F$, and any two vector spaces of cardinality $\kappa > \mathrm{card}\,F$ have bases of cardinality κ and hence are isomorphic. See also Exercises 2.32, 2.33, and 11.38.

This example is rather good in that it gives the flavour of the subject from this point. There is a huge amount of extra information to be gleaned from the algebraic arguments on vector spaces and the logical information on the theory of F-vector spaces. It is possible to classify, for example, all the possible formulas in our language of F-vector spaces, what they say and what they can define.

The example of F-vector spaces is typical in one other important regard, in that it is categorical for all cardinalities greater than the cardinality of the language. A remarkable theorem due to Morley shows that this is generally true for theories categorical in some cardinality greater than that of their language.

Theorem 11.35 (Morley) *Let T be a set of sentences in a language L and suppose that T has no finite models. Suppose further that T is κ-categorical for some cardinality $\kappa > \mathrm{card}\,L$. Then T is κ-categorical for all cardinalities $\kappa > \mathrm{card}\,L$.*

A proof of Morley's Theorem is well beyond the scope of this book. One proof goes by trying to associate a notion of *dimension* to a model of T. Then by some subtle generalisation of the argument for vector spaces a model of T is determined exactly by its dimension, which for cardinalities greater than that of the language can only be the same as the cardinality of the whole model.

Model theory says a lot about theories of dimension for rather general L-structures. In particular, a set of sentences Σ in a countable language which is κ-categorical for all uncountable κ will have the property of being ω-*stable*, and ω-stable structures have an elegant independence and dimension theory that generalise the idea of independence in a vector space. More generally, there is a slightly weaker notion of a *stable* structure which includes ω-stable structures as a special case, and stable structures also have an elegant (but not in general quite so well behaved) notion of independence. For example, the first-order theory of modules over a ring R is a stable theory with a nice notion of independence, that generalises and includes the idea of independence in a vector space over a field as a special case. *Stability theory* is a large branch of model theory that tries to analyse and classify algebraic theories of groups, rings, fields, modules, according to their properties concerning the model-theoretic notion of stability and these corresponding independence properties, and say something useful about these theories, such as what sets are definable by first-order formulas in the language.

For further information on model theory the reader should consult a more advanced text, such as the one by Marker [9].

11.2 Examples and exercises

Exercise 11.36 (a) Let A be a set of circular discs in the plane, no two of which intersect. Show that A is countable.

(b) Let B be a set of circles in the plane, no two of which intersect. Show that B may be uncountable.

(c) Let C be a set of figures-of-eight in the plane, no two of which intersect. Can C be uncountable?

Exercise 11.37 Write down the axioms for infinite F-vector spaces in the language L_F of Example 11.34. (Note that, to ensure the models of this theory are all infinite, it may be necessary to add the statements σ_n of Example 11.30.)

Exercise 11.38 Suppose V is an F-vector space of cardinality $\kappa > \operatorname{card} F$. Let B be a basis of V. Show that $\operatorname{card} B = \kappa$. What can happen when $\kappa = \operatorname{card} F$?

Definition 11.39 The theory DLO (for Dense Linear Orders) has single binary relation $<$ and axioms

- $\forall x \forall y \forall z \, (x < y \wedge y < z \rightarrow x < z)$
- $\forall x \neg (x < x)$
- $\forall x \forall y \, (x < y \vee x = y \vee y < x)$
- $\forall x \exists y \exists z \, (x < y \wedge z < x)$
- $\forall x \forall y \, (x < y \rightarrow \exists z \, (x < z \wedge z < y))$

For example, the structures $(\mathbb{Q}, \, <)$ and $(\mathbb{R}, \, <)$ both satisfy DLO.

Exercise 11.40 Prove that DLO is $\aleph_0$-categorical.

(Hint: prove that a countable model $(A, \, <) \vDash$ DLO is isomorphic to $(\mathbb{Q}, \, <)$

by building an isomorphism $A \to \mathbb{Q}$ inductively. At any stage in the construction we have n distinct elements $a_1, a_2, \ldots, a_n$ of A and n distinct elements $q_1, q_2, \ldots, q_n$ of $\mathbb{Q}$. Inductively assume $a_i < a_j$ if and only if $q_i < q_j$ for all i, j, i.e. the a_i can be ordered as $a_{i_1} < a_{i_2} < \ldots < a_{i_n}$ with the q_i ordered by the same indices, $q_{i_1} < q_{i_2} < \ldots < q_{i_n}$. Show how the DLO axioms allow, given some new a_{n+1} from A, a new element q_{n+1} to be added to the q_i list with the induction hypothesis preserved, and that, given some new q_{n+2} from $\mathbb{Q}$, a new element a_{n+2} to be added to the a_i list with the induction hypothesis preserved. (These two steps are traditionally called 'back-and-forth'.) Explain also how the countability of A and $\mathbb{Q}$, and hence enumerations of each, allow this inductive back-and-forth construction to build an isomorphism between $(A, <)$ and $(\mathbb{Q}, <)$.)

Exercise 11.41 Explain why the previous exercise shows that DLO is complete for its language.

Exercise 11.42 Find sets of sentences T_a, T_b, T_c, T_d in *countable* languages L_a, L_b, L_c, L_d such that none of T_a, T_b, T_c, T_d has any finite models, and

- T_a is κ-categorical for all infinite κ,
- T_b is κ-categorical for all infinite $\kappa > \aleph_0$ but is not $\aleph_0$-categorical,
- T_c is $\aleph_0$-categorical but is not κ-categorical for any $\kappa > \aleph_0$,
- T_d is not κ-categorical for any infinite κ.

(Hint: for T_c consider the theory DLO. Why is it not κ-categorical for $\kappa > \aleph_0$?)

A formula of the language of DLO can always be rewritten without the $\neg$ sign. For example, $\neg x = y$ is equivalent in DLO to $x < y \lor y < x$ and $\neg (x < y)$ is equivalent in DLO to $x = y \lor y < x$. We shall see soon that the quantifiers $\forall$ and $\exists$ can also be eliminated.

Definition 11.43 A formula is said to be *quantifier free* if it contains neither $\forall$ nor $\exists$.

Exercise 11.44 Using induction on the number of connectives of θ, prove that any quantifier free formula θ which only uses connectives $\lor$ and $\land$ is logically equivalent to a formula of the form

$$\bigvee_{r=1}^{k} \bigwedge_{s=1}^{l_r} \phi_{r,s}$$

where each $\phi_{r,s}$ is atomic, i.e. is an equation $t_1 = t_2$ or is $R(t_1, t_2, \ldots, t_n)$ for some relation symbol R.

A number of theories have the property that all formulas are equivalent in them to a quantifier free formula. The process showing this is called elimination of quantifiers. When a theory has elimination of quantifiers we usually have a nice description of sets that are definable in models of the theory. Quantifiers are not eliminated from the discussion completely: they are still needed in the axioms of the theory, for example. Also, from the point of view of describing the set defined by a formula, adding an existential quantifier $\exists x_n \ldots$ to a formula $\phi(x_1, \ldots, x_{n-1}, x_n)$ corresponds to a projection of the set onto the first $n-1$ coordinates. Elimination of quantifiers shows that the quantifier free definable sets are closed under projection functions, and this is often very powerful.

We will illustrate this with the theory DLO. The following lemma contains the main step in the elimination of quantifiers for DLO.

Proposition 11.45 *Suppose a formula* $\theta(x_1, \ldots, x_k)$ *of the language of DLO is of the form*

$$\exists y \bigwedge_{n=1}^{k} \phi_n(y, x_1, \ldots, x_k)$$

where each ϕ_n *is either* '$x_i < y$', '$y < x_i$', '$y = x_i$', '$x_i = x_j$', *or* '$x_i < x_j$', *for some* i, j. *Then* θ *is equivalent (in the theory DLO) to a formula which is quantifier free.*

Proof If some ϕ_n is '$y = x_i$' then we may replace y throughout with x_i and remove the quantifier. So assume that the ϕ_n involving y are all of the form '$x_i < y$', or '$y < x_i$', and let L be the set of indices of lower bounds,

$$L = \{i : \text{some } \phi_n \text{ is } x_i < y\},$$

and U be the set of indices of upper bounds,

$$U = \{i : \text{some } \phi_n \text{ is } y < x_i\}.$$

Also, let $A = \{n : \phi_n \text{ does not involve } y\}$. The more interesting case is when L and U are both non-empty, and in this case we try to rewrite θ as

$$\bigwedge_{n \in A} \phi_n \wedge \bigwedge_{i \in L} \bigwedge_{j \in U} x_i < x_j.$$

I have to explain why this statement is equivalent to θ.

The easiest way to do this is to think about a specific model of DLO with specific elements $x_1, \ldots, x_k$ in our model and argue that for these the equivalence of θ and the statement above holds. Then by the Completeness Theorem and by the fact our model and elements $x_1, \ldots, x_k$ were arbitrary we have that the equivalence is provable in DLO. But there are finitely many elements

$x_1, \ldots, x_k$ in our list, so from $\{x_i : i \in L\}$ there is a maximum element (as the order is linear), and from $\{x_i : i \in U\}$ there is a minimum element. The statement $\bigwedge_{i \in L} \bigwedge_{j \in U} x_i < x_j$ says that $\max\{x_i : i \in L\}$ is less than $\min\{x_i : i \in U\}$. So by an axiom of DLO there is some y between these. Therefore, if all the ϕ_n not involving y are also true, we deduce that $\theta(x_1, \ldots, x_n)$ is true, and this shows one direction of the equivalence of the formula above with θ. The other direction is easier since if there is y such that $x_i < y$ for all $i \in L$ and $y < x_j$ for all $i \in U$ then $x_i < x_j$ for all $i \in L$, $j \in U$ by transitivity.

In the cases when U or L is empty, we claim that θ is equivalent in DLO to $\bigwedge_{n \in A} \phi_n$, and use another axiom of DLO (the one that states there is no least or greatest element) to find a y such that y is greater than all x_i for $i \in L$ (in the case when $U = \varnothing$) or to find a y such that y is less than all x_j for $j \in U$ (in the case when $L = \varnothing$). $\square$

Exercise 11.46 Using induction on the number of quantifiers in a statement θ, and also Exercise 11.44 and Proposition 11.45, show that every formula θ in the language of DLO is equivalent in DLO to a quantifier free formula in the same language with the same free variables.

(Hint: treat the universal quantifier $\forall y \ldots$ by converting it to an existential one $\neg \exists y \neg \ldots$.)

Definition 11.47 A subset A of a model M is *definable* if there is a formula $\theta(x, y_1, \ldots, y_k)$ of the language of M and parameters $a_1, \ldots, a_k \in M$ such that

$$A = \{b \in M : M \vDash \theta(b, a_1, \ldots, a_k)\}.$$

Exercise 11.48 Show that every definable set in a model of DLO can be described as a disjoint union of singleton sets $\{a\}$ and intervals of the form $(a, b) = \{x : a < x \wedge x < b\}$ or $(a, \infty) = \{x : a < x\}$ or $(-\infty, b) = \{x : x < b\}$.

11.3 Cardinal arithmetic*

The object of this section is to provide the proofs of the various results on cardinal arithmetic that were needed above. With the exception of the Schröder–Bernstein Theorem, they all require Zorn's Lemma in some form or other. All of the results here are 'set theory' rather than logic and although the results are important for more advanced model theory they may be taken on trust.

We will start by giving Tarski's proof of the Schröder–Bernstein Theorem. The idea is that, given sets A and B and injections $f: A \to B$ and $g: B \to A$ we would like to partition $A = A_0 \cup A_1$ and $B = B_0 \cup B_1$ so that $f: A_0 \to B_0$ and

$g: B_1 \to A_1$ are onto, for then we could define $h: A \to B$ as $h(a) = f(a)$ if $a \in A_0$ and $h(a) = g^{-1}(a)$ if $a \in A_1$. To get this to work we need to find A_0, and you should note that A_0 would have to satisfy $A_0 = A \setminus g(B \setminus f(A_0))$. The following lemma is designed to provide us with such a set A_0.

Lemma 11.49 (Tarski's Fixed Point Lemma) *Let A be any set and let k be an order preserving map $k: P(K) \to P(K)$ defined on the power set $P(K) = \{U : U \subseteq K\}$ of U, where 'order preserving' means that if $X \subseteq Y \subseteq A$ then $k(X) \subseteq k(Y)$. Then there is some $A_0 \subseteq A$ such that $k(A_0) = A_0$.*

Proof The idea is to define A_0 to be the 'limit' (or more precisely, the union) of all sets $X \subseteq A$ which are too small in the sense that $X \subseteq k(X)$. More precisely, let $A_0 = \bigcup T$ where $T = \{X \subseteq A : X \subseteq k(X)\}$.

Then if $X \in T$ then $X \subseteq A_0$ so $X \subseteq k(X) \subseteq k(A_0)$. This shows that $A_0 \subseteq k(A_0)$ since each $x \in A_0$ is in some $X \in T$. Also, as k is order preserving and $A_0 \subseteq k(A_0)$ we have $k(A_0) \subseteq k(k(A_0))$ so $k(A_0) \in T$, showing that $k(A_0) \subseteq A_0$. $\square$

Proof of the Schröder–Bernstein Theorem Given sets A and B and injections $f: A \to B$ and $g: B \to A$, define a map from $k: P(A) \to P(A)$ by

$$k(X) = A \setminus g(B \setminus f(X)),$$

and verify that it is order preserving. By Tarski's Lemma there is a fixed point $A_0 \subseteq A$. Given this we may define $A_1 = A \setminus A_0$ and $h: A \to B$ as $h(a) = f(a)$ if $a \in A_0$ and $h(a) = g^{-1}(a)$ if $a \in A_1$. It is straightforward now to check that h is a bijection from A to B. $\square$

The next result shows that in almost all cases we could have defined the 'less than' operation on cardinals in terms of surjections rather than injections. (The case that would not work is that of the empty set.) Despite its apparent simplicity the Axiom of Choice or Zorn's Lemma is necessary here.

Proposition 11.50 *Suppose $f: A \to B$ is a surjection. Then $\operatorname{card} B \leqslant \operatorname{card} A$, i.e. there is an injection $g: B \to A$.*

Proof Let S be the set of all injections $g: U \to A$ such that $U \subseteq B$ and $f(g(b)) = b$ for all $b \in U$. Order S by the idea of extending functions, i.e. say that $g \leqslant h$ if and only if $\operatorname{dom} g \subseteq \operatorname{dom} h$ and $g(x) = h(x)$ for all $x \in \operatorname{dom} g$. Then the empty function is in S and S has the Zorn property as all functions in a chain of elements of S are 'compatible' so the union of this chain is also an injection from some subset of B to A satisfying the required condition.

By Zorn's Lemma there is a maximal injection $g: U \to A$ such that $f(g(b)) = b$ and we may check that $U = B$. For if not then there is $b \in B \setminus U$ and $b = f(a)$ for some $a \in A$ as f is surjective. Then we may define $h(x) = x$ if $x \in U$ and $h(b) = a$. This would be an extension of g in S, but g is maximal so there can be no such extension. Therefore $U = B$ as required.   □

If X is a non-empty set and $\sim$ an equivalence relation on X then there is always a surjection $X \to X/\sim$ given by $x \mapsto [x]$, so a quotient by an equivalence relation always has the same or smaller cardinality than the original set.

The Trichotomy Theorem is a direct consequence of the following application of Zorn's Lemma.

Proposition 11.51 *Let X, Y be sets. Then either there is an injection $X \to Y$, or there is an injection $Y \to X$.*

Proof Let S be the set of all bijections $U \to V$ where $U \subseteq X$ and $V \subseteq Y$. S is non-empty because it contains the empty function $\varnothing \to \varnothing$. Order S by the idea of extending functions, i.e. say that $f \leqslant g$ if and only if $\operatorname{dom} f \subseteq \operatorname{dom} g$ and $f(x) = g(x)$ for all $x \in \operatorname{dom} f$. Our set S has the Zorn property as all functions in a chain of elements of S are 'compatible' so the union of this chain is also a bijection from the union of the domains of the functions in the chain, to the union of their images.

By Zorn's Lemma there is a maximal bijection $f: U \to V$ where $U \subseteq X$ and $V \subseteq Y$. Since this function is maximal and cannot be extended further, there are two possibilities: either $U = X$ in which case f is an injection $X \to Y$; or $V = Y$, in which case f^{-1} is an injection $Y \to X$.   □

We now prove some results in cardinal arithmetic, starting with the more straightforward results and using these to prove more powerful results. We shall use Zorn's Lemma throughout; indeed all the remaining results in this section need Zorn's Lemma or the Axiom of Choice for their proof, including the proposition which says that the cardinality of the set of natural numbers is the least infinite cardinal. This is not quite as trivial as it may seem.

Proposition 11.52 *Let X be an infinite set, i.e. suppose that for no $n \in \mathbb{N}$ is there a bijection $f: X \to \{1, 2, \ldots, n\}$. Then $\operatorname{card} X \geqslant \operatorname{card} \mathbb{N}$, i.e. there is an injection $\mathbb{N} \to X$.*

Proof Let S be the set of finite sequences

$$(x_0, x_1, \ldots, x_{n-1})$$

of elements of X with $x_i \neq x_j$ for all $i \neq j$. We give S a partial order by defining

$$(x_0, x_1, \ldots, x_{n-1}) \leqslant (y_0, y_1, \ldots, y_{m-1})$$

to hold if and only if $n - 1 \leqslant m - 1$ and $x_i = y_i$ for all $i < n$. In other words S is ordered by the 'initial segment of' relation.

Our set S has no maximal element. This is because if $(x_0, x_1, \ldots, x_{n-1})$ is maximal then $X = \{x_0, x_1, \ldots, x_{n-1}\}$ (otherwise we could extend the sequence by at least one place), but this would contradict the assumption that X is not finite. Therefore S fails to have the Zorn property. In other words there is a chain $C \subseteq S$ with no upper bound in S. Since C is a chain, of any two sequences in C, one will always be an initial segment of the other, and since C has no upper bound in S it contains elements of arbitrarily large finite length. Thus the map $n \mapsto x_n$ where $(x_0, x_1, \ldots, x_{k-1}) \in C$ with $k > n$ defines an injection $\mathbb{N} \to X$.   □

A variation of this says essentially that $\kappa + n = \kappa$ for each infinite cardinal κ and each finite number n.

Proposition 11.53 *Let X be an infinite set and $x_0, \ldots, x_{n-1} \in X$. Then there is a bijection $X \to X \setminus \{x_0, \ldots, x_{n-1}\}$.*

Proof Assume without loss of generality that the $x_0, \ldots, x_{n-1}$ are all distinct. As in the proof of the last proposition, find an infinite sequence

$$(x_0, x_1, \ldots, x_{n-1}, x_n, \ldots)$$

of distinct elements of X by defining S to be the set of finite sequences of distinct elements of X starting with $x_0, x_1, \ldots, x_{n-1}$. Then define $f : x_i \mapsto x_{i+n}$ for x_i in this sequence and $f : x \mapsto x$ for all other $x \in X$. This is the required bijection.   □

Proposition 11.54 *Let X be an infinite set. Then there is a bijection $f : X \to (X \times \{0\}) \cup (X \times \{1\})$.*

Proof Suppose for the sake of obtaining a contradiction that there is no such function f.

Let

$$S = \{f : U \to (U \times \{0\}) \cup (U \times \{1\}) : U \subseteq X \text{ and } f \text{ is a bijection}\}.$$

This is a poset where the order relation is the notion of one function extending another, as before, and is non-empty (because S contains the empty function) and has the Zorn property (because the union of a chain of functions in S is a function with the required properties). Therefore there is a maximal element $f: U \to (U \times \{0\}) \cup (U \times \{1\})$ of S, with $U \subseteq X$. We must show that this maximal f gives a bijection $X \to (X \times \{0\}) \cup (X \times \{1\})$.

If U is in one-to-one correspondence with X, i.e. there is a bijection $g: X \to U$, then we would have a bijection $h: X \to (X \times \{0\}) \cup (X \times \{1\})$ given by $h(x) = (r, 0)$ if $f(g(x)) = (g(r), 0)$ and $h(x) = (r, 1)$ if $f(g(x)) = (g(r), 1)$. (You can check the properties.) So $X \setminus U$ must be infinite, for else if $X \setminus U = \{x_1, \ldots, x_n\}$ then by the previous proposition there would be a bijection $g: X = U \cup \{x_1, \ldots, x_n\} \to U$. Therefore there must be an injection $\mathbb{N} \to X \setminus U$, which will be notated here $i \mapsto x_i$, and using this we can extend our function f to g defined by $g(x) = f(x)$ for $x \in U$, $g(x) = (x_i, 0)$ for $x = x_{2i}$, and $g(x) = (x_i, 1)$ for $x = x_{2i+1}$. Thus f is not maximal after all. $\square$

Proposition 11.55 *Let X be infinite. Then there is a bijection $f: X \to X \times X$.*

Proof This argument is similar in structure to the last, but uses that result to obtain the contradiction to maximality. Suppose then for the sake of obtaining a contradiction that there is no such function f.

Let

$$S = \{f: U \to U \times U : U \subseteq X \text{ and } f \text{ is a bijection}\}.$$

This is made into a poset where the order relation is the notion of one function extending another, and is non-empty and has the Zorn property for the same reasons as before. Therefore there is a maximal element $f: U \to U \times U$ of S, with $U \subseteq X$. We must show that this maximal f gives a bijection $X \to X \times X$.

If U is in one-to-one correspondence with X, i.e. there is a bijection $g: X \to U$, then we would have a bijection $h: X \to X \times X$ given by $h(x) = (r, s)$ if $f(g(x)) = (g(r), g(s))$. It follows that there can be no injection $X \setminus U \to U$ for then there would be an injection $X \to U$ and hence by the Schröder–Bernstein Theorem a bijection between X and U. To see this suppose $g: X \setminus U \to U$ is an injection; then considering $X \setminus U$ via g as a subset of a second copy of U and composing maps in the obvious way we have an injection $X = U \cup (X \setminus U) \to (U \times \{0\}) \cup (U \times \{1\}) \to U$ by the previous proposition.

Since there is no injection $X \setminus U \to U$, by trichotomy there is an injection $U \to X \setminus U$, and we let V be the image of this, so $V \subseteq X \setminus U$ and V is in one-to-one correspondence with U. Then by using the previous proposition again twice together with the fact that there is a bijection $U \to U \times U$ we find a bijection

$$g: V \to (U \times V) \cup (V \times V) \cup (V \times U).$$

The union of f and g is a bijection $U \cup V \to (U \cup V) \times (U \cup V)$ contradicting the maximality of our f. □

These last two propositions and the Schröder–Bernstein Theorem easily imply the Theorem on Cardinal Arithmetic.

12

Nonstandard analysis

12.1 Infinitesimal numbers

The Completeness and Compactness Theorems for first-order logic are interesting from the point of view of the foundations of mathematics, which is what
they were originally designed for, but they also provide a powerful logical
toolkit that can be applied to other areas of mathematics. One of the most
exciting applications of the Completeness and Compactness Theorems is the
discovery by Robinson that they may be used to make perfectly rigorous sense
of the idea of an infinitesimal number, and to use infinitesimals to present
the material of traditional analysis, including continuity and differentiability.
Robinson called his method 'nonstandard analysis', which to my mind is a
somewhat unfortunate name as there is nothing at all improper about his approach. Indeed, if historical circumstances had been different, nonstandard
analysis might even have been mainstream analysis. That it is not is possibly
due to the logical difficulties some mathematicians have in understanding how
the analysis is set up – difficulties we aim to set to rights in this chapter.

Throughout this chapter I shall spell the word 'nonstandard' *without* a hy-
phen, to emphasise that this word is being used in the technical sense of 'per-
taining to infinite or infinitesimal numbers', and not in the more common
everyday sense of 'not standard' – which will never be used and always spelled
'non-standard'.

The nonstandard method involves using methods from logic to build an extended version of the real number line with infinitesimals (called the 'hyperreal
number line') and moving between the hyperreals and the usual reals. There
are several possible approaches to this, including axiomatic ways that make
the job of transferring information between the hyperreals and the reals almost
completely automatic. For workers in nonstandard analysis, these mechanical
methods of transfer are the quickest and preferred way. The approach here is

somewhat slower than many, and has been chosen to emphasise the reasoning behind it and, in particular, the use of the Compactness Theorem.

We shall discuss analysis on the reals. (Analysis on other sets can be done in exactly the same way.) We start with a first-order structure for the reals with its familiar constants, operations and relations: $(\mathbb{R}, 0, 1, +, \times, <)$. If we like, we can add other relations and functions to this structure, such as a unary (or binary) function symbol $-$ for 'subtraction' and a unary function $^{-1}$ for reciprocal. (In the case of $^{-1}$ we need to take care to define 0^{-1} since, according to the official definition of a structure, all functions must be defined on all elements of the domain. The choice of a value for 0^{-1} is entirely arbitrary; $0^{-1} = 0$ seems to be the simplest one to take.) By the usual abuse of notation, we will use the symbol $\mathbb{R}$ for both the set of reals and the structure with this domain under consideration. (If this structure and its first-order language is not clear from context it will be indicated in detail.) As the structure $\mathbb{R}$ is our main interest, we shall say that a first-order sentence σ for which $\mathbb{R} \vDash \sigma$ holds is *true*, since it is true in the real world.

Our first goal is to build a system of 'hyperreals' that looks as similar to $\mathbb{R}$ as possible but which contains infinitesimal numbers. We would like our hyperreals to contain the ordinary reals, and to achieve this we need to name all the ordinary reals with constant symbols. We introduce a constant symbol c_r for each real number $r \in \mathbb{R}$ and insist that r represents c_r itself in the structure $\mathbb{R}$. Thus we have expanded the language by adding infinitely many new symbols and saying how they are all to be interpreted. From now on, $\mathbb{R}$ will denote the corresponding structure in this expanded language. Actually, the use of the letter c and subscript r becomes cumbersome very quickly and as the constant symbol c_r always denotes the real number r we will be safe in identifying these two and using the same symbol r for each.

The set of *true sentences* in this expanded language

$$\left\{ \sigma : (\mathbb{R}, 0, 1, \ldots, r, \ldots, +, \times, -, {}^{-1}, <) \vDash \sigma \right\}$$

will be denoted $\mathrm{Th}(\mathbb{R})$.

We still do not have any infinitesimals, and this is achieved by adding yet another constant symbol, h. (I prefer to use the letter h for an infinitesimal as it is the symbol usually used in numerical analysis for a small increment, and this h and the small increment in numerical analysis play very similar roles, as we shall see.) We need to say that h is positive and smaller than any normal positive real, and this is now easy in the first-order language we have. Let

$$\Sigma = \{0 < h\} \cup \{h < r : r \in \mathbb{R}, r > 0\}.$$

This is a set of infinitely many sentences, one for each positive $r \in \mathbb{R}$. We want to show that $\text{Th}(\mathbb{R}) \cup \Sigma$ has a model, and to do this we use the Compactness Theorem.

Let $\Sigma_0 \subseteq \Sigma$ be finite. Then there are finitely many sentences $h < r$ in Σ_0 and hence amongst these sentences there is one, $h < r_0$, where r_0 is minimum. Clearly $r_0 > 0$ (since otherwise $h < r_0$ would not be in Σ), so we may temporarily interpret h as $r_0/2$. In so doing we make a new structure $(\mathbb{R}, \ldots, h)$ for the language of $\Sigma \cup \text{Th}(\mathbb{R})$ and as this h is positive and less than all r appearing in Σ_0, we have

$$(\mathbb{R}, 0, 1, \ldots, r, \ldots, +, \times, -, {}^{-1}, <, h) \vDash \Sigma_0 \cup \text{Th}(\mathbb{R}).$$

This applies to any $\Sigma_0 \subseteq \Sigma$, though the choice of the interpretation of h will differ for different Σ_0. Thus the conditions of the Compactness Theorem are met, and the conclusion is that there is a new structure $^*\mathbb{R}$ that makes all the statements in $\Sigma \cup \text{Th}(\mathbb{R})$ true simultaneously. That is,

$$^*\mathbb{R} = (^*\mathbb{R}, 0, 1, \ldots, r, \ldots, +, \times, -, {}^{-1}, <, h) \vDash \Sigma \cup \text{Th}(\mathbb{R})$$

where we identify the name of the structure with its underlying set, as usual.

The Compactness Theorem provided us with this new structure, but gives us little extra information about it that we did not have already in the sentences from $\Sigma \cup \text{Th}(\mathbb{R})$. The procedure to find out anything about $^*\mathbb{R}$ is to write down sentences from $\text{Th}(\mathbb{R})$ and use the fact that they are also true in $^*\mathbb{R}$. For example, if r, s are distinct real numbers then they represent distinct values of $^*\mathbb{R}$. This is because the sentence $\neg(r = s)$ is true in $\mathbb{R}$ and hence in $^*\mathbb{R}$. Therefore we may safely continue to identify r with the value it represents in $^*\mathbb{R}$; in other words we may regard $^*\mathbb{R}$ as a *superset* of $\mathbb{R}$. This embedding of $\mathbb{R}$ into $^*\mathbb{R}$ preserves all the structure in $\mathbb{R}$, for if $r + s = t$ is true, then this is a sentence in $\text{Th}(\mathbb{R})$ and hence is true in $^*\mathbb{R}$ too. The same argument applies to all the other function and relation symbols of the language. Thus $\mathbb{R}$ is a *subfield* of $^*\mathbb{R}$, and we are justified in thinking of $^*\mathbb{R}$ as an extension of the number line $\mathbb{R}$ that also contains infinitesimals such as h.

The hyperreals, $^*\mathbb{R}$, contains many infinitesimals, such as

$$3h, 2h, h/2, h/3, h^2, h^3, \ldots$$

To see that these are all different, it is necessary to write down true first-order statements again. For example,

$$\forall x \, (0 < x \rightarrow \neg(x = 2 \times x))$$

is true, hence true in $^*\mathbb{R}$, and shows that $h \neq 2h$. As well as containing infinitesimal numbers, $^*\mathbb{R}$ also contains infinite numbers such as h^{-1}. That this is greater than all standard reals is because of the sentences

$$0 < r \to 0 < r^{-1}$$

and

$$\forall x (0 < x \wedge x < r^{-1} \to r < x^{-1})$$

which are both true for all positive $r \in \mathbb{R}$. There are of course infinitely many infinite numbers in $^*\mathbb{R}$ for similar reasons, and near each standard real $r \in \mathbb{R}$ there are hyperreals such as $r + h$ and $r - h$ which are 'infinitesimally close' to r.

To further develop our picture of $^*\mathbb{R}$ and to draw some conclusions about first-order logic, it may be helpful to consider the Archimedean Property of $\mathbb{R}$. The *Archimedean Property* for an ordered field F states that

- For all $x \in F$ there is $n \in \mathbb{N}$ such that $x < 1 + 1 + \cdots + 1$ where 1 appears n times.

This is not true for our hyperreals, as infinite numbers such as h^{-1} are not bounded above by natural numbers. Thus there is no statement in $\mathrm{Th}(\mathbb{R})$ which is equivalent in both $\mathbb{R}$ and $^*\mathbb{R}$ to the Archimedean Property, because if there were, it would be true, hence hold in $^*\mathbb{R}$ and $^*\mathbb{R}$ would be Archimedean. Put another way: the Archimedean Property is not first-order. If we examine the statement above we see the cause of the problem. The expression 'there is a natural number n such that ...' cannot be rewritten using quantifiers $\forall x \ldots$ and $\exists y \ldots$ where the variables range over elements of the field in question.

A similar problem occurs for another familiar property of the reals, that for every bounded subset there is a least upper bound. $\mathbb{R}$ itself is a bounded subset of $^*\mathbb{R}$ (bounded above and below by h^{-1} and $-h^{-1}$) but has no least upper bound, since if $x \in {}^*\mathbb{R}$ is infinite then $x - 1 \in {}^*\mathbb{R}$ would also be infinite (exercise). This property of $\mathbb{R}$ (confusingly for us also called *completeness of*, though it has nothing to do with the Completeness Theorem of logic) is therefore also not first-order. Once again, the offending phrase is a quantification over objects that are not elements of the field, in this case 'for every bounded subset ...'.

To see how the hyperreals can be used to define derivatives, etc., consider the function $f : x \mapsto x^2$. Suppose $x \in \mathbb{R}$ and h is an infinitesimal (any positive or negative infinitesimal, not necessarily our original h). Then we have

$$\frac{f(x+h) - f(x)}{h} = \frac{(x+h) \times (x+h) - x \times x}{h} = \frac{2xh + h^2}{h} = 2x + h.$$

This is because these facts are all represented by true first-order statements starting $\forall x \forall h \, (h \neq 0 \rightarrow \dots)$ hence true in ${}^*\mathbb{R}$. Thus $(f(x+h) - f(x))/h$ is $2x$ plus an infinitesimal, and this should be used to conclude that the derivative of $f(x)$ is $2x$.

To make this precise we need a way of 'disregarding an infinitesimal' like h in $2x + h$.

Definition 12.1 A hyperreal $x \in {}^*\mathbb{R}$ is *finite* if it is bounded above and below by standard reals. That is, x is finite if there are $r, s \in \mathbb{R}$ such that $s < x < r$. The set of finite hyperreals is denoted ${}^*\mathbb{R}_{\mathrm{fin}}$.

If $x \in {}^*\mathbb{R}$ is finite, then the set $\{r \in \mathbb{R} : r < x\}$ is non-empty and bounded above. Sets of reals A like this which are non-empty and bounded above have a *least upper bound* or *supremum* denoted sup$A$.

Definition 12.2 If $x \in {}^*\mathbb{R}$ is finite, we define its *standard part*, st(x) by

$$\mathrm{st}(x) = \sup \{r \in \mathbb{R} : r < x\}.$$

Thus the standard part operation is a map st: ${}^*\mathbb{R}_{\mathrm{fin}} \rightarrow \mathbb{R}$, taking finite hyperreals to the standard reals. For any $r \in \mathbb{R}$ and any (positive or negative) infinitesimal h, we have st$(r+h) = r$, since st$(x) = \sup \{s \in \mathbb{R} : s < r+h\}$ which is $\sup \{s \in \mathbb{R} : s \leqslant r\} = r$ if $h > 0$ and $\sup \{s \in \mathbb{R} : s < r\} = r$ if $h \leqslant 0$. Since there are many infinitesimals, this means that many hyperreals get sent to the same standard part.

Using this idea we can say the derivative of a function like $f(x) = x^2$ should be $2x$ since

$$\mathrm{st}((f(x+h) - f(x))/h) = 2x$$

for all $x \in \mathbb{R}$ and all infinitesimals $h \neq 0$.

12.2 Examples and exercises

Exercise 12.3 Prove that for any $r \in \mathbb{R}$ and any (positive or negative) infinitesimal h, we have st$(r+h) = r$.

Exercise 12.4 Let h be infinitesimal. Prove that st $\frac{1}{1+h} = 1$.

Exercise 12.5 Explain why the relation $<$ in ${}^*\mathbb{R}$ is a linear order.

Exercise 12.6 Show that there is no first-order formula $\theta(x)$ such that, for all $a \in {}^*\mathbb{R}$,

$${}^*\mathbb{R} \vDash \theta(a) \text{ if and only if } a \text{ is infinitesimal.}$$

(Hint: assuming such a formula exists, write down a sentence of $\mathrm{Th}(\mathbb{R})$ that shows $\mathbb{R} \vDash \exists x\, \theta(x)$ and deduce that some $r \neq 0$ in $\mathbb{R}$ satisfies $\theta(r)$ in ${}^*\mathbb{R}$.)

Exercise 12.7 Prove in detail the assertion made above that if $h \in {}^*\mathbb{R}$ is infinitesimal and positive and $r \in \mathbb{R}$ is positive then $r \times h \in {}^*\mathbb{R}$ is infinitesimal and, for $r, s \in \mathbb{R}$, $r \times h = s \times h$ implies $r = s$.

Exercise 12.8 Let $h \in {}^*\mathbb{R}$ be a positive infinitesimal. Show that $h^{-1} > n$ for all $n \in \mathbb{N}$.

Exercise 12.9 Show that the standard part map st: ${}^*\mathbb{R}_{\mathrm{fin}} \to \mathbb{R}$ preserves $+$, $\times$ and $\leqslant$. That is, if $x, y, z \in {}^*\mathbb{R}_{\mathrm{fin}}$ we have: $x + y = z$ implies $\mathrm{st}(x) + \mathrm{st}(y) = \mathrm{st}(z)$; $x \times y = z$ implies $\mathrm{st}(x) \times \mathrm{st}(y) = \mathrm{st}(z)$; and $x \leqslant y$ implies $\mathrm{st}(x) \leqslant \mathrm{st}(y)$. Explain why similar statements for the strict order $<$ and reciprocal $^{-1}$ fail.

12.3 Overspill and applications*

We are now going to develop the ideas presented in the previous section and apply them to give some applications for real analysis. The discussion above was comparatively simple – simplistic perhaps – and our first task is to extend the ideas to more difficult problems than differentiating functions like x^2.

Our discussion in the last section relied on x^2 being definable in the language. In fact, the language we took for the reals was rather limited and very few functions we might be interested in are definable there. The solution is to start from a richer structure for the reals with more functions.

In the approach I am taking here, there are some choices to be made concerning the structure to study, and in particular what functions and relations it should have. In some sense any nonstandard approach to analysis seems to require some choice of structure at the outset, but the process is not nearly as bad as it sounds, and there are some good ways to get round this issue which more or less guarantee that all the functions you are interested in will be there. (See Lindstrøm's article in Cutland [5] for example.) I want to emphasise the precise links with the Completeness Theorem, so my approach will be somewhat more pedestrian.

For the rest of this section we shall take as our structure $\mathbb{R}$ with all possible functions $f\colon \mathbb{R}^k \to \mathbb{R}$ and all possible relations $R \subseteq \mathbb{R}^l$ $(k, l \in \mathbb{N})$. This gives a structure for set of

$$\mathbb{R} = (\mathbb{R}, \ldots, r, \ldots, R, \ldots, f, \ldots)$$

for a huge uncountable language (where constant symbols r, relation symbols R, and function symbols f are identified with the real number, set or function they represent), but it has the advantage that all functions (such as sin, cos, exp) and sets (such as $\mathbb{N}$, $\mathbb{Q}$) are represented. Of course the usual arithmetic functions such as $+$, $\times$, etc., and the usual relations such as $<$ are represented just as before.

Let $\mathrm{Th}(\mathbb{R})$ be the set of sentences true in this structure. By the Compactness Theorem there is a structure $*\mathbb{R} \supseteq \mathbb{R}$ for the same language, containing infinitesimals and making every sentence in $\mathrm{Th}(\mathbb{R})$ true. The usual notation for the functions and relations in $*\mathbb{R}$ corresponding to $f\colon \mathbb{R}^k \to \mathbb{R}$ and $R \subseteq \mathbb{R}^l$ is to add a star, so these are notated $*f\colon *\mathbb{R}^k \to *\mathbb{R}$ and $*R \subseteq *\mathbb{R}^l$. The set $*\mathbb{R}$ is often called the set of hyperreals.

Of course $\mathbb{N} \subseteq \mathbb{R}$, so there is a relation symbol for the naturals in $\mathbb{R}$. We start by considering its nonstandard version $*\mathbb{N} \subseteq *\mathbb{R}$.

Adding new relations as we did makes new properties first-order. For instance the Archimedean Property is now a first-order statement about $\mathbb{R}$,

$$\forall x \in \mathbb{R} \, \exists n \in \mathbb{N} \, x < n,$$

or, as variables range over real numbers anyway, we may as well write

$$\forall x \, \exists n \, (n \in \mathbb{N} \wedge x < n).$$

Thus this statement is in $\mathrm{Th}(\mathbb{R})$ and the corresponding version is true for the hyperreals:

$$\forall x \in *\mathbb{R} \, \exists n \in *\mathbb{N} \, x < n.$$

The hyperreal field $*\mathbb{R}$ is not however Archimedean: it still contains infinitesimals and infinite numbers just as before. Instead, the statement above shows that for each infinite number such as h^{-1} there are elements $n \in *\mathbb{N}$ greater than it. This is such an important fact we shall record it here as a proposition.

Proposition 12.10 *The nonstandard version $*\mathbb{N}$ of the set of natural numbers contains infinite numbers. The finite elements of $*\mathbb{N}$ are exactly the standard naturals $n \in \mathbb{N}$.*

Proof The first part has already been explained. For the second, suppose $x \in *\mathbb{N} \wedge x < r$ where $r \in \mathbb{R}$. Then let $n \in \mathbb{N}$ be the largest integer such that

$n < r$. So $\forall x \, (x < r \wedge x \in \mathbb{N} \rightarrow x = 0 \vee x = 1 \vee \ldots \vee x = n)$ is a true first-order statement, hence true in the nonstandard world. It follows x is equal to one of the standard natural numbers $0, 1, \ldots, n$.   □

The main tool for nonstandard analysis is the overspill principle, which in its simplest form says that the set of standard natural numbers cannot be given by a first-order formula in the structure $*\mathbb{R}$.

Definition 12.11 A set $A \subseteq *\mathbb{R}^n$ is *parameter-definable* in $*\mathbb{R}$ if there is a formula $\theta(x_1, \ldots, x_n, y_1, \ldots, y_k)$ in the free-variables shown and hyperreals $r_1, \ldots, r_k \in *\mathbb{R}$ such that

$$A = \left\{ (x_1, \ldots, x_n) \in *\mathbb{R}^n : *\mathbb{R} \vDash \theta(x_1, \ldots, z_n, r_1, \ldots, r_k) \right\}.$$

A function $f : *\mathbb{R}^n \rightarrow *\mathbb{R}$ is *parameter-definable* in $*\mathbb{R}$ if the set

$$\left\{ (x_1, \ldots, x_n, x_{n+1}) \in *\mathbb{R}^{n+1} : f(x_1, \ldots, x_n) = x_{n+1} \right\}$$

is parameter-definable in $*\mathbb{R}$. Parameter-definable sets and functions will also be called *internal*.

Theorem 12.12 (Overspill) *The set* $\mathbb{N}$ *is not parameter-definable in* $*\mathbb{R}$. *In fact, if a set* $A \subset *\mathbb{R}$ *is parameter-definable in* $*\mathbb{R}$ *and contains all standard natural numbers* $n \in \mathbb{N}$ *then it must also contain some infinite natural numbers.*

Proof Suppose $\theta(x, y_1, \ldots, y_n)$ and $r_1, \ldots, r_n \subset *\mathbb{R}$ are such that

$$\mathbb{N} = \{x \in *\mathbb{N} : *\mathbb{R} \vDash \theta(x, r_1, \ldots, r_n)\}.$$

Then $*\mathbb{R} \vDash \theta(0, r_1, \ldots, r_n)$ and also

$$*\mathbb{R} \vdash \forall x \, (x \in *\mathbb{N} \wedge \theta(x, r_1, \ldots, r_n) \rightarrow \theta(x+1, r_1, \ldots, r_n))$$

since $x \in *\mathbb{N}$ and $\theta(x, r_1, \ldots, r_n)$ implies $x \in \mathbb{N}$ hence $x + 1 \in \mathbb{N}$. But then it would follow from the true statement

$$\forall \bar{y} \, (\theta(0, \bar{y}) \wedge \forall x \, (x \in \mathbb{N} \wedge \theta(x, \bar{y}) \rightarrow \theta(x+1, \bar{y})) \rightarrow \forall x \, (x \in \mathbb{N} \rightarrow \theta(x, \bar{y})))$$

(where $\bar{y}$ denotes $y_1, \ldots, y_n$) that

$$*\mathbb{R} \vDash \forall x \, (x \in *\mathbb{N} \rightarrow \theta(x, r_1, \ldots, r_n)),$$

or in other words

$$*\mathbb{N} = \{x \in *\mathbb{N} : *\mathbb{R} \vDash \theta(x, r_1, \ldots, r_n)\}.$$

But $*\mathbb{N}$ contains infinite numbers, so this contradicts our assumption that the formula $\theta(x, r_1, \ldots, r_n)$ is only true for finite natural numbers x.   □

The scope of this book prevents us from giving many applications of nonstandard analysis. Those that we do give will use $^*\mathbb{R}$ to define new functions on $\mathbb{R}$. The procedure typically is to take an internal function $f\colon {}^*\mathbb{R}\to{}^*\mathbb{R}$ definable in the first-order language (with parameters) such that the restriction of f to the finite hyperreals maps into the finite hyperreals: $f\restriction{}^*\mathbb{R}_{\mathrm{fin}}\colon {}^*\mathbb{R}_{\mathrm{fin}}\to{}^*\mathbb{R}_{\mathrm{fin}}$. We may then define a new function $\mathrm{st}\,f\colon \mathbb{R}\to\mathbb{R}$ by $(\mathrm{st}\,f)(x)=\mathrm{st}(f(x))$, for $x\in\mathbb{R}$. Overspill is the main tool used to prove properties of $\mathrm{st}\,f$ such as continuity, differentiability. As a by-product our theory also gives elegant alternative characterisations of continuity, differentiability, etc., for standard functions.

We start by looking at continuity. For a function defined as above, the next theorem says that $\mathrm{st}\,f$ is continuous at a if and only if $f(y)$ is infinitesimally close to $f(a)$ whenever y is infinitesimally close to a.

Theorem 12.13 *Suppose* $f\colon {}^*\mathbb{R}\to{}^*\mathbb{R}$ *is internal,* $a\in\mathbb{R}$, *and* $f\restriction{}^*\mathbb{R}_{\mathrm{fin}}\colon {}^*\mathbb{R}_{\mathrm{fin}}\to{}^*\mathbb{R}_{\mathrm{fin}}$. *Suppose also that*

$$\textit{for all } y\in{}^*\mathbb{R},\ \ \mathrm{st}(y)=a \textit{ implies } \mathrm{st}(f(y))=\mathrm{st}(f(a)).$$

Then the function $\mathrm{st}\,f\colon \mathbb{R}\to\mathbb{R}$ *defined by* $(\mathrm{st}\,f)(x)=\mathrm{st}(f(x))$ *is continuous at* a.

Proof Suppose $\mathrm{st}(y)=a$ implies $\mathrm{st}(f(y))=\mathrm{st}(f(a))$ for all $y\in{}^*\mathbb{R}$, and let $\varepsilon>0$ be an arbitrary standard real. We look at the internal set

$$A=\left\{n\in{}^*\mathbb{N}:\exists y\in{}^*\mathbb{R}\left(|y-a|>\frac{1}{n+1}\wedge|f(y)-f(a)|\geqslant\varepsilon\right)\right\}.$$

If $A\supseteq\mathbb{N}$ then by overspill there is an infinite $n\in{}^*\mathbb{N}$ in A and hence some y in the interval $(a-\frac{1}{n+1},a+\frac{1}{n+1})$ with $|f(y)-f(a)|\geqslant\varepsilon$. This would mean $\mathrm{st}(y)=a$ and $\mathrm{st}(f(y))\neq\mathrm{st}(f(a))$, contradicting the assumption. So instead there must be some standard $n\in\mathbb{N}$ in A for which the statement

$$\forall y\in{}^*\mathbb{R}\left(|y-a|\leqslant\frac{1}{n+1}\to|f(y)-f(a)|<\varepsilon\right)$$

is true in $^*\mathbb{R}$. It follows that for all standard $y\in\mathbb{R}$ with $y\in(a-\frac{1}{n+1},a+\frac{1}{n+1})$ we have $f(a)-\varepsilon<f(y)<f(a)+\varepsilon$ true in $^*\mathbb{R}$ so

$$\mathrm{st}(f(a))-\varepsilon\leqslant\mathrm{st}(f(y))\leqslant\mathrm{st}(f(a))+\varepsilon,$$

which suffices to show $\mathrm{st}\,f$ is continuous at a. $\square$

Every standard function $f\colon \mathbb{R}\to\mathbb{R}$ has a nonstandard version $^*f\colon {}^*\mathbb{R}\to{}^*\mathbb{R}$; also, for each $x\in\mathbb{R}$ and $y=f(x)\in\mathbb{R}$ the statement $y=f(x)$ is true and first-order, so $y={}^*f(x)$ holds in the nonstandard world. Thus $f(x)=\mathrm{st}(^*f(x))$ for all $x\in\mathbb{R}$, so the above result gives a useful criterion for when a function f

is continuous at $a \in \mathbb{R}$. In fact this criterion is exact, as the following result shows.

Theorem 12.14 *A function* $f : \mathbb{R} \to \mathbb{R}$ *is continuous at* a *if and only if for all* $y \in {}^*\mathbb{R}$, $\mathrm{st}(y) = a$ *implies* $\mathrm{st}({}^*f(y)) = a$.

Proof One direction has already been proved. For the other, suppose f is continuous at a, so

$$\forall \varepsilon > 0 \, \exists \delta > 0 \, \forall h < \delta \, |f(a+h) - f(a)| < \varepsilon.$$

Let $\varepsilon > 0$ be a standard real number and choose another standard real number $\delta > 0$ such that

$$\forall h < \delta \, |f(a+h) - f(a)| < \varepsilon.$$

This last statement about ε, δ is first-order hence true in ${}^*\mathbb{R}$, so

$$\forall h < \delta \, |{}^*f(a+h) - {}^*f(a)| < \varepsilon.$$

So in particular $|{}^*f(a+h) - {}^*f(a)| < \varepsilon$ holds for any infinitesimal h. But ε was arbitrary, so $|{}^*f(a+h) - {}^*f(a)|$ is also infinitesimal, hence

$$\mathrm{st}({}^*f(a+h)) = f(a)$$

for any infinitesimal h. □

We can define differentiation in a similar way.

Theorem 12.15 *Suppose* $f : {}^*\mathbb{R} \to {}^*\mathbb{R}$ *is internal, suppose* $a, b \in \mathbb{R}$, *and that* $f \upharpoonright {}^*\mathbb{R}_{\mathrm{fin}} : {}^*\mathbb{R}_{\mathrm{fin}} \to {}^*\mathbb{R}_{\mathrm{fin}}$. *Suppose also that*

$$\mathrm{st}\left(\frac{f(a+h) - f(a)}{h} \right) = b$$

for all non-zero infinitesimals $h \in {}^*\mathbb{R}$. *Then the function* $\mathrm{st} f$ *is differentiable at* a *with derivative* b.

Proof Let $\varepsilon > 0$ be a standard real. We will find a positive $n \in \mathbb{N}$ such that for all standard $h \in \mathbb{R}$ with $0 < |h| < 1/n$,

$$\left| \frac{\mathrm{st} f(a+h) - \mathrm{st} f(a)}{h} - b \right| \leqslant \varepsilon.$$

Note that as the difference of two infinitesimals is infinitesimal and h is standard and positive $(\operatorname{st} f(a+h) - \operatorname{st} f(a))/h = \operatorname{st}((f(a+h) - f(a))/h)$. It therefore suffices to find a positive $n \in \mathbb{N}$ such that

$$\forall h \left(0 < |h| < 1/n \to \left| \frac{f(a+h) - f(a)}{h} - b \right| < \varepsilon \right)$$

holds in $^{*}\mathbb{R}$. But by assumption, the statement just given is first-order and true for all infinite $n \in {}^{*}\mathbb{N}$. If it is false for all finite $n \in {}^{*}\mathbb{N}$ then the formula

$$n \in {}^{*}\mathbb{N} \wedge \exists h \left(0 < |h| < 1/n \wedge \left| \frac{f(a+h) - f(a)}{h} - b \right| \geqslant \varepsilon \right)$$

would define the set $\mathbb{N}$ in $^{*}\mathbb{R}$, which is impossible by overspill. Therefore such a finite $n \in \mathbb{N}$ exists as required. $\qquad\square$

This idea can be used to give a nonstandard characterisation of differentiability.

Theorem 12.16 *Suppose* $f \colon \mathbb{R} \to \mathbb{R}$, *and* $a, b \in \mathbb{R}$. *Then the function* f *is differentiable at* a *with derivative* b *if and only if*

$$\operatorname{st} \left(\frac{{}^{*}f(a+h) - {}^{*}f(a)}{h} \right) = b$$

for all infinitesimals $h \in {}^{*}\mathbb{R}$.

Proof Left as an exercise. $\qquad\square$

We are finally in a position to prove a more substantial result: we will prove the *Peano Existence Theorem* for a solution to a first-order differential equation, and the argument will be based on the classical Euler method for its numerical solution.

The *Euler method* for solving a differential equation $y'(t) = F(y(t), t)$ numerically, subject to a given value for $y(0)$, asks us to choose a small stepsize h and define a sequence of points (x_n, y_n) approximating the solution by $x_0 = 0$, $y_0 = y(0)$, $x_{n+1} = x_n + h$, and $y_{n+1} = y_n + hF(x_n, y_n)$. Notice that this approximates the differential equation with a 'difference equation' $y_{n+1} - y_n = hF(nh, y_n)$. The connections between differential equations and difference equations go very deep.

Theorem 12.17 (Peano Existence Theorem) *Let* $F \colon \mathbb{R} \times [0, 1] \to \mathbb{R}$ *be a continuous function, which is bounded, i.e. there is* $B \in \mathbb{R}$ *such that* $|F(t, y)| < B$ *for all* t, y. *Suppose also* $y_0 \in \mathbb{R}$. *Then there is a continuous and differentiable function* $y \colon [0, 1] \to \mathbb{R}$ *such that* $y(0) = y_0$ *and* $y'(t) = F(y(t), t)$ *for all* $t \in [0, 1]$.

Proof Using the idea of the Euler method, we inductively define for each positive $h \in \mathbb{R}$ a function $Y_h(t)$ so that

$$Y_h(t) = y_0 \text{ if } t \in [0, h)$$

and

$$Y_h(t) = Y_h(t - h) + hF(([t/h] - 1)h, Y_h(t - h)) \text{ if } t \in [h, 1]$$

where $[t/h]$ denotes the integer part of t/h. Thus $Y_h(t)$ is just a stepped version of the Euler method solution defined on the whole of $[0, 1]$. (Using the simpler looking $t - h$ instead of $([t/h] - 1)h$ would give a slightly different function, but one that would work just as well for the proof that follows.)

There is no difficulty about the existence of the function Y_h. It is defined using induction in approximately $1/h$ steps. In fact, we can regard $Y_h(t)$ as a function of two arguments, writing it as $Y(h, t)$. This will enable us to use the function $^*Y(h, t)$ in our nonstandard universe.

Now let $h \in {}^*\mathbb{R}$ be a positive infinitesimal and define $y(t) = \text{st}(^*Y(h, t))$. We need to show that this is well defined (i.e. that $^*Y(h, t)$ is finite for all $t \in [0, 1]$) and differentiable with derivative satisfying $y'(t) = F(y(t), t)$.

To do this, use the fact that F is bounded and there is $B \in \mathbb{R}$ such that $|F(t, y)| < B$ for all t, y. Thus $|Y_h(t_1) - Y_h(t_2)| < B|t_1 - t_2|$ for all t_1, t_2 and h, since at most $|t_1 - t_2|/h$ steps are required in going from t_1 to t_2 and each one adds or subtracts less than hB. This is a first-order statement about $Y_h(t) = Y(h, t)$ so transfers to $^*Y(h, t)$:

$$\forall h > 0 \, \forall t_1, t_2 \in [0, 1] \; |^*Y(h, t_1) - {}^*Y(h, t_2)| < B|t_1 - t_2|.$$

It follows from $^*Y(h, 0) = y_0$ that $y_0 - B < {}^*Y(h, t) < y_0 + B$ for $t \in [0, 1]$ and hence $^*Y(h, t)$ is always finite and $y(t)$ is well defined.

We use Theorem 12.15 to show that $y(t)$ is differentiable and find its derivative. Let $t \in [0, 1]$ be standard and $k \in {}^*\mathbb{R}$ a non-zero infinitesimal. For simplicity assume $k > 0$; the case when $k < 0$ is identical with signs reversed. We need to estimate $^*Y(h, t + k) - {}^*Y(h, t)$. This number is the sum of $[k/h]$ terms of the form $h \times {}^*F(t', {}^*Y(h, t'))$ where $t \leqslant t' \leqslant t + k$. For such t', note that $|t - t'| \leqslant k$ is infinitesimal and we can employ the formula above bounding $|^*Y(h, t_1) - {}^*Y(h, t_2)|$ to see that

$$\left|^*Y(h, t) - {}^*Y(h, t')\right| < B|t - t'|$$

is also infinitesimal. Thus $^*F(t', {}^*Y(h, t')) = {}^*F(t + r, {}^*Y(h, t) + s)$ for some infinitesimals r, s and therefore $|{}^*F(t', {}^*Y(h, t')) - {}^*F(t, {}^*Y(h, t))|$ is infinitesimal by Theorem 12.14 and the continuity of F. Putting all this together,

$$\frac{{}^*Y(h, t+k) - {}^*Y(h, t)}{k} = \frac{[k/h] \times h \times ({}^*F(t, {}^*Y(h, t)) + \text{infinitesimal})}{k}$$

which is ${}^*F(t, {}^*Y(h, t)) + \text{infinitesimal}$. We conclude that $y(t)$ is differentiable with derivative $y'(t) = \text{st}({}^*F(t, {}^*Y(h, t))) = F(t, y(t))$, by the continuity of F again. $\square$

We are going to conclude with one further example of nonstandard analysis: its use in constructing a function $f: [0, 1] \to [0, 1]$ which is everywhere-continuous and nowhere-differentiable. The idea once again is to define a family of functions $f_n(x)$ in $\mathbb{R}$ indexed by $n \in \mathbb{N}$ and then consider the function $f_n(x)$ in ${}^*\mathbb{R}$ where $n \in {}^*\mathbb{N}$ is infinite. This use of an infinite n replaces the usual limit process of classical analysis.

We define, for all $x \in [0, 1]$, f_0 by

$$f_0(x) = 0$$

and f_1 by

$$f_1(x) = f_0(x) + (x - 1/4) \text{ for all } x \in [1/4, 1/2],$$

and

$$f_1(x) = f_0(x) + (3/4 - x) \text{ for all } x \in [1/2, 3/4],$$

where $f_1(x) = f_0(x)$ for all other x. More generally, for $n \in \mathbb{N}$, we define f_{n+1} by

$$f_{n+1}(x) = f_n(x) + (x - a) \text{ for all } x \in [a, (a+b)/2],$$

and

$$f_{n+1}(x) = f_0(x) + (b - x) \text{ for all } x \in [(a+b)/2, b],$$

for all $[a, b]$ with a, b given by $a = (4k+1)/4^{n+1}$ and $b = (4k+3)/4^{n+1}$, for $k \in \mathbb{N}$, setting $f_{n+1}(x) = f_n(x)$ for all other x. Figure 12.1 shows the first four functions in this sequence.

Once again, there is no problem in defining $f_n(x)$ for each $n \in \mathbb{N}$ by this induction. In fact, we may consider $f_n(x)$ as a two-argument function $f(n, x)$ as soon as we make some arbitrary choice of $f(y, x)$ for $y \notin \mathbb{N}$ such as $f(y, x) = 0$ for such y.

One helpful fact about this family of functions is that at the 'corner' points where $x = i/4^n$ ($i \in \mathbb{N}$, $0 \leqslant i \leqslant 4^n$) the value $f_m(x)$ is fixed for sufficiently large

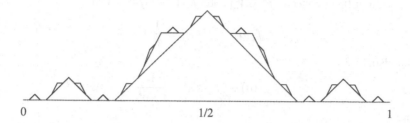

0   1/2   1

Figure 12.1 The construction of a nowhere-differentiable function.

m: in fact, $f_m(x) = f_n(x)$ for all $m \geqslant n$. This fact can be written as a first-order statement about $\mathbb{R}$:

$$\forall n \in \mathbb{N} \, \forall i \in \mathbb{N} \, (0 \leqslant i \leqslant 4^n \to \forall m \in \mathbb{N} \, (m \geqslant n \to f_m(i/4^n) = f_n(i/4^n))).$$

Another useful fact about these functions is that f_n and f_{n+1} do not differ by very much. Indeed, checking the definition one can see that

$$|f_{n+1}(x) - f_n(x)| \leqslant \frac{1}{4^{n+1}}$$

for all $x \subset [0, 1]$ and hence for all $k \geqslant n$

$$|f_k(x) - f_n(x)| \leqslant \frac{1}{4^{n+1}} + \frac{1}{4^{n+2}} + \cdots + \frac{1}{4^k} < \frac{1}{4^{n+1}}\left(1 + \frac{1}{4} + \cdots\right) - \frac{1}{3 \times 4^n}.$$

In particular, $f_k(x) < 1/3$ for all k and all x. Once again these facts may be written down in a first-order way and transferred to $^*\mathbb{R}$.

We now define the actual function we are interested in. It is the function $f = \operatorname{st}^* f_n$ where $^*f_n(x) = {}^*f(n, x)$ in $^*\mathbb{R}$ and $n \in {}^*\mathbb{N}$ is some fixed infinite natural number. Thus, for $x \in [0, 1]$, $f(x) = \operatorname{st}(^*f_n(x))$.

Theorem 12.18 *The function $f\colon [0, 1] \to \mathbb{R}$ is continuous at all $a \in [0, 1]$ and differentiable at no $a \in [0, 1]$.*

Proof We prove continuity by using Theorem 12.14. Let $a \in [0, 1]$ and $x \in {}^*[0, 1]$ be infinitesimally close to a. Suppose $a < x$, the other case being argued in an almost identical way, and take $i \in {}^*\mathbb{N}$ and $k \in {}^*\mathbb{N}$ such that

$$\frac{i}{4^k} \leqslant a < x \leqslant \frac{i+1}{4^k}.$$

Since $\operatorname{st}(x) = a$ we may take $k \leqslant n$ to be an infinite natural number here.

Then $^*f_n(i/4^k) = {}^*f_k(i/4^k)$, and $^*f_n((i+1)/4^k) = {}^*f_k((i+1)/4^k)$, and $^*f_k$ is a straight line between $^*f_k(i/4^k)$ and $^*f_k((i+1)/4^k)$. Thus

$$|{}^*f_k(a) - {}^*f_k(x)| < \frac{1}{4^k}.$$

But also,

$$|{}^*f_k(a) - {}^*f_n(a)| < \frac{1}{3 \times 4^k}$$

and

$$|{}^*f_k(x) - {}^*f_n(x)| < \frac{1}{3 \times 4^k}$$

giving

$$|{}^*f_n(a) - {}^*f_n(x)| < \frac{5}{3 \times 4^k}$$

which is infinitesimal. Hence f is continuous at a.

To see that f is not differentiable at any $a \in [0, 1]$ we use the classical definition of differentiability. Let $a \in [0, 1]$ and let $\delta > 0$ be an arbitrary positive standard real number. Choose $i, k \in \mathbb{N}$ such that $i/4^k \leqslant a \leqslant (i+1)/4^k$ with $1/4^k < \delta$, and write x_1 for $i/4^k$ and x_2 for $(i+1)/4^k$. We are going to look at f_k, f_{k+1} and f on the interval $[x_1, x_2]$.

Note first that for all $x \in [x_1, x_2]$ we have $0 \leqslant f_k(x) \leqslant f_{k+1}(x) \leqslant f(x)$ with $|f(x) - f_{k+1}(x)| \leqslant 1/(3 \times 4^{k+1})$. The interval $[x_1, x_2]$ may be divided into four parts, as $[u_0, u_1], [u_1, u_2], [u_2, u_3], [u_3, u_4]$, where $u_0 = x_1$, $u_1 = x_1 + 1/4^{k+1}$, $u_2 = x_1 + 2/4^{k+1}$, $u_3 = x_1 + 3/4^{k+1}$, and $u_2 = x_1 + 4/4^{k+1} = x_2$. We also have $f(u) = f_{k+1}(u)$ when u is an endpoint u_i of one of these intervals. Note also that $f(u_2) = f_{k+1}(u_2) = f_k(u_2) + 1/4^{k+1}$ and that f_k is a straight line between u_0 and u_4. Let $v_i = f_k(u_i)$ for each i.

We suppose first that $u_0 \leqslant a \leqslant u_1$ and $v_0 \leqslant v_4$ and we shall estimate the gradients of the chords from $(a, f(a))$ to $(u_2, f(u_2))$ and from $(a, f(a))$ to $(u_4, f(u_4))$. Set $t = (a - u_0)/(u_4 - u_0)$, the proportion of $[u_0, u_4]$ covered by $[u_0, a]$. Then for the first chord, we have

$$f(u_2) - f(a) \geqslant \left(f_k(u_2) + \frac{1}{4^{k+1}} \right) - \left(f_k(a) + \frac{1}{3 \times 4^{k+1}} \right).$$

But f_k is a straight line on this interval, so

$$f(u_2) - f(a) \geqslant \frac{1}{4^{k+1}} + \frac{1}{2}(v_0 + v_4) - (1-t)v_0 - tv_4 - \frac{1}{3 \times 4^{k+1}}$$

hence

$$f(u_2) - f(a) \geqslant \left(\frac{1}{2} - t \right)(v_4 - v_0) + 4^{-(k+1)} \times \frac{2}{3}.$$

The horizontal distance from a to u_2 is $(1/2-t)/4^k$ and so the gradient of this chord from $(a, f(a))$ to $(u_2, f(u_2))$ is at least

$$4^k(v_4 - v_0) + \frac{1}{6}\left(\frac{1}{2} - t\right)^{-1}.$$

As $t \geqslant 0$ this is at least

$$4^k(v_4 - v_0) + \frac{1}{3}.$$

The gradient of the chord from $(a, f(a))$ to $(u_4, f(u_4))$ is a little easier to estimate. As $f(a) \geqslant f_k(a)$ it is at most

$$\frac{f_k(u_4) - f_k(a)}{(1-t) \times 4^{-k}} = \frac{v_4 - (1-t)v_0 - tv_4}{(1-t) \times 4^{-k}} = \frac{(1-t)(v_4 - v_0)}{(1-t) \times 4^{-k}} = 4^k(v_4 - v_0).$$

Thus the difference between these two gradients is at least $1/3$, which is a positive number, independent of k and hence independent of δ. Therefore the function f is not differentiable at a since for each δ we can find two chords within distance δ of a that differ in gradient by $1/3$.

The cases for the other values for a, between u_1 and u_2, etc., and when $v_4 < v_0$ are treated with a similar calculation of gradients of suitably chosen chords, and are left for the reader to complete and verify. In all cases the conclusion is that f is not differentiable at a, as we require.   □

Overspill, as we have seen, is an important tool for nonstandard analysis. In its simplest form it is, however, only a statement about how the standard natural numbers sit inside the set of nonstandard natural numbers. As such, it is also a tool for understanding number theory, $\text{Th}(\mathbb{N}, 0, 1, +, \times, <)$ say. Thus in principle nonstandard techniques might be applied to number theory, though it turns out that there are (at present) fewer convincing applications in this area. On the other hand, nonstandard models of number theory (or arithmetic, as they are more usually called) are interesting objects with links and applications to many other areas, especially model theory and algebra. For an introduction to models of arithmetic, see Kaye [7].

Exercise 12.19 Show that $^*\mathbb{N}$ contains infinitely many infinite prime numbers.

Exercise 12.20 Suppose that $^*\mathbb{N}$ contains an infinite prime number p such that $p+2$ is also prime. Show that this would imply that $\mathbb{N}$ contains infinitely many primes p such that $p+2$ is also prime.

The last exercise shows how nonstandard methods may be of use in number theory too: proving the existence of a single nonstandard number with a given property implies the existence of infinitely many standard numbers with the same property.

Exercise 12.21 Let $f: \mathbb{R} \to \mathbb{R}$ and $g: \mathbb{R} \to \mathbb{R}$ be functions and $a \in \mathbb{R}$. Suppose f and g are continuous at a and $g(a) \neq 0$. Use nonstandard methods to show $f(x)/g(x)$ is continuous at a (when considered as a function of x). Contrast your proof with a traditional epsilon-delta proof.

Exercise 12.22 Prove Theorem 12.16 giving the nonstandard characterisation of differentiability.

Exercise 12.23 Prove the chain rule, that if g is differentiable at $a \in \mathbb{R}$ and f is differentiable at $g(a) \in \mathbb{R}$ then the composition $f \circ g$ is differentiable at a with derivative $(f \circ g)'(a) = f'(g(a)) \times g'(a)$. (Hint: work in a hyperreal structure with versions $^*f$ and $^*g$ of both f, g.)

References

[1] Martin Aigner and Günter M. Ziegler, *Proofs from The Book*. Berlin: Springer-Verlag, third edition, 2004.

[2] George S. Boolos and Richard C. Jeffrey, *Computability and Logic*. Cambridge: Cambridge University Press, 1989.

[3] Lewis Carroll, *Symbolic logic: Part II. Advanced*, edited with annotations and an introduction by William Warren Bartley. New York: Clarkson N. Potter, 1977.

[4] Nigel Cutland, *Computability*. Cambridge: Cambridge University Press, 1980.

[5] Nigel Cutland, editor, *Nonstandard Analysis and its Applications*, Volume 10 of *London Mathematical Society Student Texts*. Cambridge: Cambridge University Press, 1988.

[6] Douglas R. Hofstadter, *Gödel, Escher, Bach: an Eternal Golden Braid*. New York: Basic Books, 1979.

[7] Richard Kaye, *Models of Peano Arithmetic*, Volume 15 of *Oxford Logic Guides*. Oxford: Oxford University Press, 1991.

[8] Azriel Lévy, *Basic Set Theory*. Berlin: Springer-Verlag, 1979. Reprinted Mineola, NY: Dover Publications, 2002.

[9] David Marker, *Model Theory: an Introduction*, Volume 217 of *Graduate Texts in Mathematics*. New York: Springer-Verlag, 2002.

[10] Thomas L. Saaty and Paul C. Kainen, *The Four-Color Problem: Assaults and Conquest*. New York: Dover Publications, second edition, 1986.

[11] Stephen G. Simpson, editor, *Reverse Mathematics 2001*, Volume 21 of *Lecture Notes in Logic*. La Jolla, CA: Association for Symbolic Logic, 2005.

[12] Robin Wilson, *Four Colors Suffice: How the Map Problem Was Solved*. Princeton, NJ: Princeton University Press, 2002.

Index

Printed in the United States
By Bookmasters